Create Stunning Renders Using V-Ray in 3ds Max

Create Stunning Renders Using V-Ray in 3ds Max

Guiding the Next Generation of 3D Renderers

Margarita Nikita

CRC Press
Taylor & Francis Group
Boca Raton London New York

CRC Press is an imprint of the
Taylor & Francis Group, an **informa** business

AN A K PETERS BOOK

First edition published 2022
by CRC Press
6000 Broken Sound Parkway NW, Suite 300, Boca Raton, FL 33487-2742

and by CRC Press
2 Park Square, Milton Park, Abingdon, Oxon, OX14 4RN

© 2022 Margarita Nikita

CRC Press is an imprint of Taylor & Francis Group, LLC

Reasonable efforts have been made to publish reliable data and information, but the author and publisher cannot assume responsibility for the validity of all materials or the consequences of their use. The authors and publishers have attempted to trace the copyright holders of all material reproduced in this publication and apologize to copyright holders if permission to publish in this form has not been obtained. If any copyright material has not been acknowledged please write and let us know so we may rectify in any future reprint.

Except as permitted under U.S. Copyright Law, no part of this book may be reprinted, reproduced, transmitted, or utilized in any form by any electronic, mechanical, or other means, now known or hereafter invented, including photocopying, microfilming, and recording, or in any information storage or retrieval system, without written permission from the publishers.

For permission to photocopy or use material electronically from this work, access www.copyright.com or contact the Copyright Clearance Center, Inc. (CCC), 222 Rosewood Drive, Danvers, MA 01923, 978-750-8400. For works that are not available on CCC please contact mpkbookspermissions@tandf.co.uk

Trademark Notice: Product or corporate names may be trademarks or registered trademarks and are used only for identification and explanation without intent to infringe.

Library of Congress Cataloging-in-Publication Data
Names: Nikita, Margarita, author.
Title: Create stunning renders using V-ray in 3ds max : guiding the next generation of 3D renderers / Margarita Nikita.
Description: First edition. | Boca Raton : CRC Press, 2022. | Includes bibliographical references and index.
Identifiers: LCCN 2021022801 | ISBN 9780367701482 (hardback) | ISBN 9780367701352 (paperback) | ISBN 9781003144786 (ebook)
Subjects: LCSH: V-ray. | 3ds max (Computer file) | Architectural rendering—Computer-aided design.
Classification: LCC NA2728 .N55 2022 | DDC 720.28/40285536—dc23
LC record available at https://lccn.loc.gov/2021022801

ISBN: 978-0-367-70148-2 (hbk)
ISBN: 978-0-367-70135-2 (pbk)
ISBN: 978-1-003-14478-6 (ebk)

DOI: 10.1201/9781003144786

Typeset in Minion
by KnowledgeWorks Global Ltd.

Access the Support Material: http://routledge.com/9780367701482

To John

Contents

Acknowledgments		xi
My Story		xiii
Author Bio		xv

1 Getting Started — 1

- 1.1 3ds Max Interface Overview 1
- 1.2 Viewports 4
- 1.3 Create and Modify Standard Primitives 8
- 1.4 Selecting Objects 11
- 1.5 Transform Commands 12
  - 1.5.1 Select and Move 12
  - 1.5.2 Select and Rotate 13
  - 1.5.3 Select and Scale 14
- 1.6 Saving a Project 16
  - 1.6.1 Save As 16
  - 1.6.2 Save Selected 16
  - 1.6.3 Archive 16

2 Rendering Settings — 19

- 2.1 Assign V-Ray 19
- 2.2 V-Ray Frame Buffer 20
  - 2.2.1 Overview 20
    - 2.2.1.1 Rendering Window 22
    - 2.2.1.2 V-Ray Messages Window 23
    - 2.2.1.3 VFB Window 23
  - 2.2.2 Layers 23
    - 2.2.2.1 Exposure 24
    - 2.2.2.2 White Balance 26

vii

		2.2.2.3	Curves 27
		2.2.2.4	Lens Effects 29
	2.2.3	Saving a Render 32	
	2.2.4	History 32	
2.3	Rendering Settings...37		
	2.3.1	V-Ray..37	
		2.3.1.1	Global Switches 37
		2.3.1.2	Image Sampler (Antialiasing) 39
		2.3.1.3	Progressive Image Sampler............................. 42
		2.3.1.4	Bucket Image Sampler..................................... 45
	2.3.2	GI 49	
		2.3.2.1	Global Illumination 49
	2.3.3	Render Elements... 50	
	2.3.4	Common..51	
2.4	Interactive Production Rendering (IPR)..................................57		

3 Cameras   61

3.1	Placing a VRayPhysicalCamera...61	
	3.1.1	Selecting a Camera... 63
	3.1.2	Using Transforms to Adjust the Camera Position 65
3.2	VRayPhysicalCamera Settings... 67	
	3.2.1	Focal Length 68
	3.2.2	Clipping Planes... 68
	3.2.3	Using the Camera Viewport Controls to Adjust the Camera Position 72
	3.2.4	Simulation of Real-Life Cameras by VRayPhysicalCamera 73
3.3	Camera Lister 83	

4 Natural Lighting   85

4.1	VRaySun 85	
	4.1.1	Placing VRaySun.. 85
	4.1.2	VRaySun Parameters... 88
4.2	HDRIs and Dome VRayLight..91	
	4.2.1	Dome VRayLight Settings 95

5 Artificial Lighting   101

5.1	VRayLights 101
5.2	VRayIES 116
5.3	VRayLightMix... 123

6 Material Editor   127

6.1	Slate Material Editor... 127
6.2	Compact Material Editor... 138
6.3	Slate vs Compact Material Editor.. 144

7 Materials   147

- 7.1 The Scene Materials 147
  - 7.1.1 Fabric 149
  - 7.1.2 Wood 151
  - 7.1.3 Glass 157
  - 7.1.4 Metal 159
  - 7.1.5 Drapery and Sheer 161
  - 7.1.6 Leather 163
  - 7.1.7 Carpet 168
  - 7.1.8 Book (Multi/Sub-Object Material) 169
  - 7.1.9 Greenery (Multi/Sub-Object Material) 173
- 7.2 Asset Tracking 178
- 7.3 UVW Map 180

8 Libraries   191

- 8.1 VRayMtl Presets 191
- 8.2 V-Ray Material Library 191
- 8.3 Cosmos Browser 197
- 8.4 Bedroom Transformation 201

Index   207

Acknowledgments

I would like to thank Alena Ivakina and Marianna Sazina, my team members in High Q Renders, for 3D modeling the objects used in the book's exercise. Special thanks to my family for always supporting me.

My Story

My first years as a renderer remind me of my first day at the gym; I went there with my hopes up and after 10 minutes on the treadmill, I was feeling exhausted and disappointed.

3D rendering is like muscles; you need to train them daily, and no matter how hard it seems at the beginning, if you are committed, it becomes so much easier. For those wondering, I still suck at the gym, but my commitment to rendering has definitely paid off.

Fifteen years later, I am the co-founder of High Q Renders LLC, an award-winning 3D rendering studio based in San Francisco, CA, and with offices in Greece. We are actively collaborating with some of the most renowned buildings and hotel brands worldwide, such as the Marriott Hotels & Resorts, the Hilton Hotels, the Ritz Carlton Luxury Residences, and the Hyatt. Our clientele includes celebrity interior designers Kari Whitman and Fox Nahem with very prestigious clients like President Obama, Jessica Alba, Melanie Griffith, and Robert Downey Jr.

My passion for teaching is equal to my enthusiasm for making renders and that is why I wrote this book. This book is intended for architects, interior designers, and anyone else wanting to create photorealistic renderings using V-Ray in 3ds Max. You do not need to have any experience to follow this book, but any prior knowledge of working in 3ds Max will help you jump right in.

This book covers only 3D rendering and does not describe any modeling techniques. More specifically, you will learn the basics of 3ds Max, how to assign V-Ray, and how to adjust your rendering settings. Then, you will see how to add cameras and set up the natural and artificial lighting in your scene. Finally, you will learn how to apply materials to a scene and how to use the libraries that come with V-Ray. All the above are taught using as an example a bedroom scene, which you can download and follow the steps. To download the exercise, visit the book's website. The link is available on the first page of each chapter (DOI). The versions used and described are 3ds Max 2020 and V-Ray 5.

Author Bio

Margarita Nikita is the co-founder of High Q Renders LLC, an award-winning creative company based in San Francisco, CA, with offices in Greece. She also runs a training center in her native country, Greece, on 2D and 3D design software. She has published several design books, some of which are used in university courses, actively contributing to the formation of the new generation of 3D renderers in Greece. She shares her knowledge, advice, and tips and tricks on her YouTube channel, *Margarita Nikita*. More of her work is available at her Instagram account, *@margarita.nikita*.

1

Getting Started

V-Ray is a plugin that can be used in several design software like *3ds Max, Sketchup, Maya,* etc. In this book, we will explore the use of *V-Ray* in *3ds Max 2020.* If this is your first experience with *3ds Max,* it is useful to get familiar first with the interface, before getting into the rendering process. Thus, in this chapter, the basic features of the *3ds Max* interface are presented. This chapter concludes with the creation and modification of standard primitives as well as the description of the transform commands, i.e., the commands related to the selection, movement, rotation, and scale of an object since these commands are used extensively in the upcoming chapters. Finally, the different ways that a project can be saved, depending on whether it will be stored or shared, are described.

1.1 3ds Max Interface Overview

When you open *Autodesk 3ds Max 2020,* a *Welcome Screen* appears. It is a set of slides designed to provide new users with basic information to help them get started (Figure 1.1).

When you close the *Welcome Screen,* you see the *3ds Max* interface (Figure 1.2), which consists of the following items:

1. **Title bar:** Shows the name of the project and the version of the *3ds Max.* Every new project is named by default *Untitled* and you need to *Save* the project to rename it (See Section 1.6).
2. **Menu bar (File, Edit, Tools, …):** Contains drop-down menus with commands. The name of each menu indicates the purpose of the commands.
3. **Main Toolbar:** Provides quick access to some of the most commonly used commands in *3ds Max,* like the commands *Undo, Redo, Move, Rotate,* or *Scale.*

DOI: 10.1201/9781003144786-1

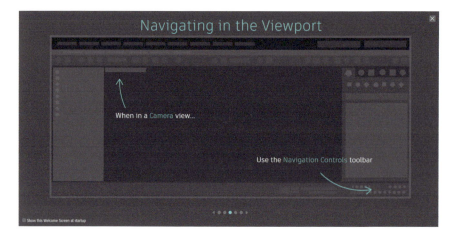

Figure 1.1
Welcome Screen.

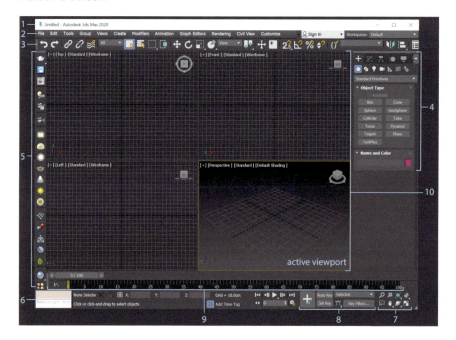

Figure 1.2
3ds Max interface.

4. **Command Panel:** Consists of six sub-panels: *Create, Modify, Hierarchy, Motion, Display,* and *Utilities.* They include controls for creating objects, editing them, animation and display options, and miscellaneous utilities. You use this panel mainly to create and edit the cameras and the lights in a scene.

5. **V-Ray Toolbar:** Contains shortcuts for some of the most commonly used V-Ray commands.
6. **Status Bar:** On the left side there is a two-line interface, where you can create scripts and execute commands. On its right, there is the *Status line,* which displays the number and type of object(s) selected, and below the *Status line,* there is the *Prompt line,* giving instructions on what your next step should be. On the right of the *Status line,* there is the *Coordinate Display* area with the X, Y, and Z fields indicating the coordinates of the selected object and allowing you to control its position (see Section 1.5).
7. **Viewport Navigation Controls:** Includes buttons that control the display and navigation of the viewports. Some of the buttons change depending on which viewport is active. See more details in Chapter 3.
8. **Animation and Time Controls:** Contains the main controls for animation. However, this book will not cover any animation techniques.
9. **Time Slider:** Allows you to move through any frame of the animation.
10. **Viewports:** Everything in *3ds Max* is located in a three-dimensional world that is viewed through one or more (up to 4) viewports. By using multiple viewports, you can have the best possible visualization of objects in a scene (see Sections 1.2 and 1.5).

If you look at the various toolbars, you will notice a double dotted line at the beginning of each toolbar. If you move the cursor there, it changes appearance to a double-cross, and you can click and drag to reposition the toolbar in the *3ds Max* interface (Figure 1.3).

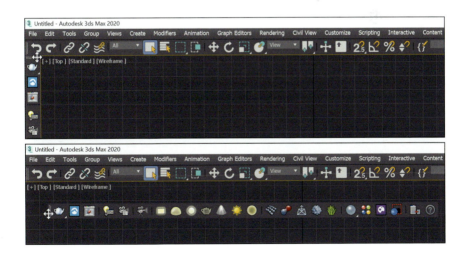

Figure 1.3

Repositioning the *V-Ray Toolbar*.

1.1 3ds Max Interface Overview

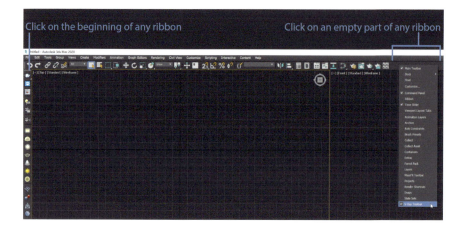

Figure 1.4
Ways to hide/unhide a toolbar.

To hide or unhide a toolbar, right-click on the double dotted line at the beginning of a toolbar or on an empty area at any toolbar's ribbon and select/deselect from the pop-up menu any toolbar (Figure 1.4).

1.2 Viewports

When you open *3ds Max 2020*, there are by default four viewports displayed—*Top, Front, Left,* and *Perspective.* The multiple viewports help to observe different aspects of the scene. One of the viewports is marked with a highlighted border and is the active viewport, as seen in Figure 1.1. To make a viewport active, simply click on it. Press *Alt+W* to maximize the active viewport or to switch from one viewport back to the four viewports. Another way to perform this action is to click on the *Maximize Viewport Toggle,* , from the *Viewport Navigation Controls* at the bottom right corner of the work environment (Figure 1.5).

To resize the viewports, drag the intersection of two or four viewports. To return to the original layout, right-click on an intersection and choose *Reset Layout* (Figure 1.6).

Figure 1.5
Maximize Viewport Toggle command.

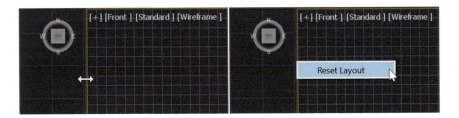

Figure 1.6

Control the size of the layout from the intersection of the viewports.

A viewport can be rotated; hold down the *Alt* key and drag in a viewport using the middle mouse button. Otherwise, use the *Orbit* command from the *Navigation Controls* (Figure 1.7). Note that this command rotates the viewport and, therefore, it rotates all objects in the scene that exist in that viewport. The other viewports remain still. If you want to rotate a specific object in a scene, you should use the command *Select and Rotate*, described in Section 1.5. In this case, the object is rotated in all viewports.

To change the number of viewports and their layout, go to the menu *Views* and choose *Viewport Configuration*.... Otherwise, in any viewport, click the *General* viewport label, [+], and choose *Configure Viewports* from the pop-up menu. Click on the *Layout* tab and choose a layout. Assign what each viewport will display by clicking on the viewport representation and choosing from the pop-up menu. Click *OK* for the changes to apply (Figure 1.8).

Revert to the default viewport layout. In the lower-left corner of each viewport, a three-color world-space tripod is visible. The colors correspond to the three axes of world space: red for X, green for Y, and blue for Z. In the upper-right corner of each viewport is the *ViewCube*. This tool shows the orientation of the scene based on the North direction of the model. Each viewport contains a series of four menus in the top left corner that control the viewport display. The menus are displayed with brackets and from left to right are: *General, Point-of-View (POV), Shading*, and *Per View Preference* (Figure 1.9).

Figure 1.7

Orbit command.

1.2 Viewports

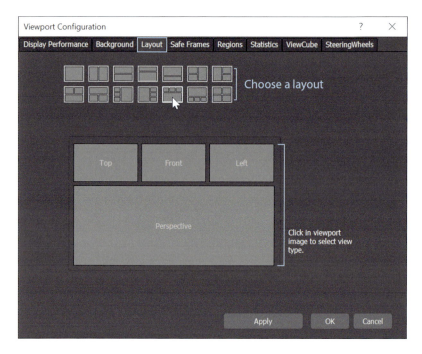

Figure 1.8
Viewport Configuration.

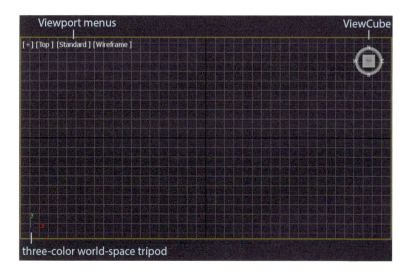

Figure 1.9
Viewport menus, *ViewCube*, and three-color world-space tripod.

- **General [+]:** Use this menu to control the viewport display and configuration. Some useful commands are the *Maximize viewport* to maximize/minimize the selected viewport and the *Active Viewport* command to choose which viewport will be activated. Two other useful commands in the *General* menu are the *Grid* and the *Configure Viewports...* Use the *Grid* to hide or unhide the viewport grid and the *Configure Viewports...* to choose a different layout for the viewports via the *Layout* tab. The *Grid* command can also be executed by pressing the *G* key.
- **POV:** Use this menu to change the active point of view. That is, the position from which we see the scene. Thus, some basic options are *Top, Front, Left, Perspective, Cameras*, where the last option concerns the view of a scene via a camera.
- **Shading:** Use this menu to set the shading of the viewport. The default option is *Standard*, which applies standard-quality settings for shading and lighting. If, for instance, you choose *Performance*, the scene will be displayed with a medium grey color. This means that any textures applied will no longer be visible in the viewport, but the scene will render with the textures. This option can be used in scenes with many polygons to maximize the performance of the viewport.
- **Per View Preference:** This last menu controls the way an object is displayed in the viewports. Two are the basic options, *Wireframe* and *Default Shading*. *Wireframe* means that the objects will be displayed as wired objects, while *Default Shading* means that the objects will be displayed with their colors and textures. Another useful option is the *Edged Faces*. In simple terms, this is a combination of *Wireframe* and *Default Shading*. The objects are displayed with the color/texture, whereas the edges are also highlighted.

As an example, Figure 1.10 shows a box in two *Perspective* viewports using *Wireframe* (left) and *Default Shading* (right) display.

Figure 1.10

Perspective viewports displaying the box in two different ways.

1.2 Viewports　　　　　　　　　　　　　　　　　　　　　　　　　　　　　　7

1.3 Create and Modify Standard Primitives

In this section, we will create some basic primitives and edit them. We will start with a box and continue with a sphere. You should keep in mind that when we do modeling in *3ds Max*, the first thing we need to do is to set up the units. Therefore, go to the menu *Customize* and choose *Units Setup*. Choose a unit system, for example, *Metric* and from the drop-down list choose the unit scale, say *Centimeters*. It is also important to click on the *System Unit Setup* button and set the same units, otherwise, you will not be able to measure correctly (Figure 1.11).

> Be careful to set the units from the very beginning when you start the project and not after you have created the geometry of the project.

To create a basic primitive, go to the *Command Panel*, click on the first tab, *Create*, ➕, and below there is a row of categories you can create. You can choose between *Geometry*, *Shapes*, *Lights*, *Cameras*, *Helpers*, *Space Warps*, and *Systems*. Click on the first button, *Geometry*, ◯. From the drop-down menu, make sure *Standard Primitives* is selected and click on *Box* from the *Object Type* rollout. Go to the *Top* viewport, click anywhere, and drag to create a rectangle, which is

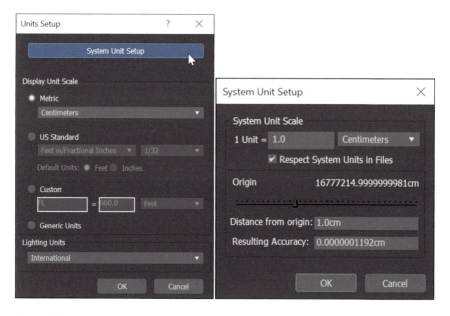

Figure 1.11

System Unit Setup.

8   1. Getting Started

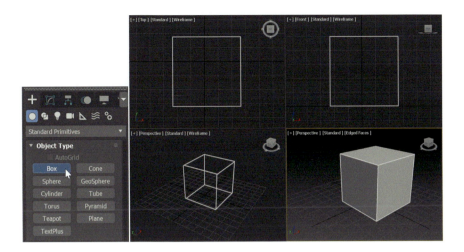

Figure 1.12
Box creation.

the base of the box. Release the cursor to complete the base, then move the mouse to give height to the box and click to complete the design process (Figure 1.12).

The box you created does not have exact dimensions. To specify them, go to the *Modify* tab, , the second button in the *Command panel*. If you want the box to be 50 cm on each side, go to the *Length*, *Width*, and *Height* fields and type 50.

To see the dimensions of an object in the *Modify* tab, the object must be selected.

Go to the *Navigation Controls* at the bottom right corner of the work environment and click on the *Zoom Extends All Selected* command. This command zooms the selected object in the viewports (Figure 1.13).

Figure 1.13
Zoom Extends All Selected command.

1.3 Create and Modify Standard Primitives

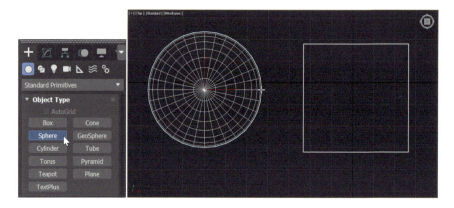

Figure 1.14
Sphere creation.

Create another standard primitive, a sphere. Go to the *Top* viewport and zoom out using the middle mouse button to create some space for the sphere. By holding down the middle mouse button, the *Pan* command is enabled. Move the viewport so that the box is moved to the right. Go to the *Create* tab and choose *Sphere*. In the *Top* viewport, click and drag to define the radius. When you release the mouse, the sphere is created (Figure 1.14).

As before, the sphere does not have an exact radius. To specify the radius, go to the *Modify* tab and type 25 cm. If you click on the *Zoom Extends All Selected* command, the sphere will get centered in all viewports. If you want both the box and the sphere to get centered, then click and hold down the *Zoom Extends All Selected* command and choose *Zoom Extends All*. Now all the objects zoom in and not only the selected ones (Figure 1.15).

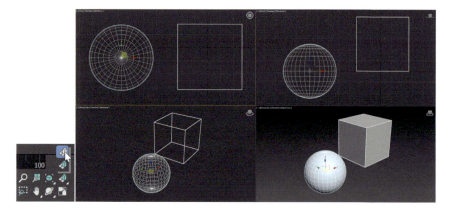

Figure 1.15
All objects centered in the viewports using the *Zoom Extends All* command.

Figure 1.16
Select an Object command.

1.4 Selecting Objects

To select an object, click on the *Select an Object* command, ![icon], from the *Main Toolbar* and then click on the object you want to select (Figure 1.16). If you want to select more than one object, hold down the *Ctrl* key.

Another way is to click on the *Select by Name* command, ![icon]. The *Select From Scene* dialog box appears that shows all the objects of the scene in alphabetical order. If you click on the *Box001* and press *OK*, the box is selected (Figure 1.17).

To open the *Select From Scene* dialog box, you can also press the *H* key. At the top of the dialog box, you see the same icons that are in the *Create* tab in the *Command Panel*. They are used to narrow down your research. This is a helpful feature, because in the final scene, you may have hundreds of objects, many cameras, and several lights, so when you narrow down the research by category, it is easier to find what you are looking for. When the icon has a blue background, this category is enabled, and its elements are visible in the list.

When an object is selected, its edges acquire a white color, so that you can easily identify it from the other objects of the scene. Also, a x, y, and z coordination system appears on it.

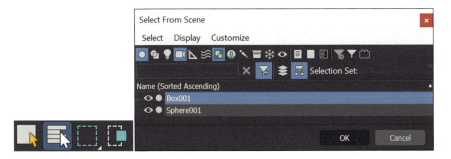

Figure 1.17
Select by Name command (left) and *Select From Scene* dialog box (right).

When you create something in *3ds Max*, it is automatically named by the software. To see the name, select the object and check the *Name and Color* field in the *Command Panel*. To rename it, type in the desired name. Next to the name field is a color swatch. This color is not the material of the object, but the 2D color representation of the object. If you click on the color field and choose a yellow color, the box edges will appear yellow in the viewports.

Go to the *Top* viewport and select the *Box001*. If you click in another viewport, for example in the *Perspective,* the viewport is activated, but the object is deselected. To activate a viewport without losing the selection, you must right-click in the new viewport.

1.5 Transform Commands

The transform commands are the *Select and Move, Select and Rotate,* and *Select and Scale* commands. Although this book only covers rendering techniques and not modeling, these commands are crucial to prepare a scene for rendering and that is the reason we present them.

1.5.1 Select and Move

As its name denotes, use this command to select and move objects. To activate the command, go to the *Main Toolbar* and click on the *Select and Move* button, ⊕ or press *W* (Figure 1.18).

When you activate the *Select and Move* command a x, y, z coordination system is placed on the selected object. Depending on the arrow you click on and drag, the movement is constrained to the respective axis. If, for instance, you click on the red arrow, which represents the x axis, and drag the cursor, the object moves only along the x axis. Apart from the red, green, and blue arrows, there are also some small planes that connect the axes at the origin. The edges of those planes normally have the color of the axis they touch on. These planes are used to move an object along two axes simultaneously (Figure 1.19).

To move an object by a specific distance, use the *Coordinates Display* in the *Status Bar*. If, for instance, you want to move the box 50 cm to the right, enable the *Select and Move* command, select the box, and go to the X field. These fields show the coordinates of the box. If you click on the button *Absolute Mode Transform Type-In,* ▣, in front of the X, Y, Z fields, the fields change to zero. Type 50 in the *X* field, press *Enter,* and the box moves to the right by 50 cm. Once you press *Enter,* the value changes again to 0. The name and appearance of the *Absolute Mode Transform Type-In* button changed to *Offset Mode Transform Type-In,* ▣

Figure 1.18

Select and Move command from the *Main Toolbar.*

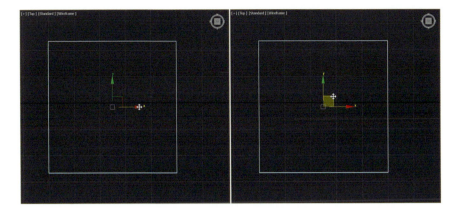

Figure 1.19

Box movement using the arrows or the small plane.

Figure 1.20

Absolute Mode (left) and *Offset Mode* (right) *Transform Type-In*.

(Figure 1.20). Click on it to return to the default display of the coordinates of the selected object.

1.5.2 Select and Rotate

To rotate an object, use the *Select and Rotate* command, , from the *Main Toolbar* or press the *E* key (Figure 1.21).

When this command is enabled, in the *Perspective* viewport you see three colored circles and an external white one at the center of the selected object, which define the rotation axis. In the other viewports (*Top, Bottom, Left, Right,* …), the three colored circles, due to the projection, are reduced to one colored circle and a colored cross.

If you click on a circle and drag the cursor up or down, you rotate the object along that circle. Use the *Transform Type-In* fields to set a specific orbit of the selected object. If, for instance, you type 45 in the Z field, the box will rotate by 45 degrees clockwise around the z axis (Figure 1.22).

Figure 1.21

Select and Rotate command from the *Main Toolbar.*

1.5 Transform Commands

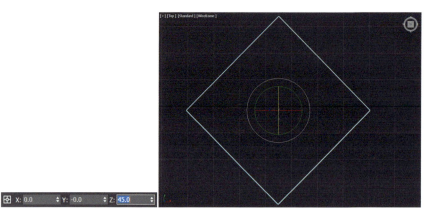

Figure 1.22

Box rotation by 45° using the Z *Transform Type-In*.

1.5.3 Select and Scale

To scale an object, use the *Select and Scale* command, from the *Main Toolbar* (Figure 1.23).

As an example, go to the *Top* viewport, enable the *Select and Scale* command and choose the box. Click on the red arrow and drag to scale the box along the x axis, while if you click on the green arrow and drag, you scale it along the y axis. Apart from the two axes, there are also two triangles that connect the two axes, a highlighted yellow and an outlined yellow. Click and drag on the highlighted yellow to scale the box along the x, y, and z axis simultaneously. If you click and drag on the outlined yellow, then you scale the box along the x and y axis simultaneously, but the size along the z axis remains the same (Figure 1.24).

To scale the selected object by a specific percentage, for example, 50%, use the *Absolute Mode Transform Type-In fields*. With the *Top* viewport active, go to the *X* field, type 50, and press *Enter*. The box is half the size.

> It is useful to check the *Perspective* viewport while transforming an object, to fully understand the changes you are making.

A point that needs special attention regarding the *Scale* command is the following. In the previous example, the dimensions of the box were scaled down by 50%. However, if you go to the *Modify* tab, to check the dimensions of this box, you will

Figure 1.23

Select and Scale command from the *Main Toolbar*.

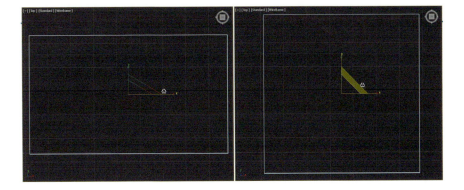

Figure 1.24

Box scaling using the arrows or the triangles.

notice that they are still 50 cm, while they should be 25 cm. Unfortunately, these fields are not updated when using the *Select* and *Scale* command. For this reason, when you want to check the dimensions of a selected object, it is preferable not to use the *Modify* tab, but the *Utilities* tab, 🔧, in the *Command Panel*. Thus, select the object (*Box001*), go to the *Command Panel*, and from the *Utilities* tab click on *Measure*. Go to the *Dimensions* section to see the dimensions of the selected object (Figure 1.25).

Figure 1.25

Measure command and its use to determine the dimensions of the selected object.

1.5 Transform Commands

1.6 Saving a Project

When you start working on a new project, it is automatically named *Untitled* by the software. To rename it, you need to save it. To save a project, go to the menu *File* and choose *Save* or press *Ctl+S*. In the *Save File As* dialog box, type in the desired name, and choose the destination folder. If you press again *Ctl+S*, the *Save File As* dialog box does not appear since you have already named the project and you just save the progress of the file.

I strongly advise you, while you are working on a project to keep saving it often, especially when you work with heavy files, with many polygons, so that you will not have any unpleasant surprises; for instance, *3ds Max* may crash and you lose your progress.

1.6.1 Save As

The *Save As* command allows you to save a project under a different name, in a different location, or in a different version. For example, go to the menu *File* and choose *Save As*. The *Save File As* dialog box appears. From the *Save type as* field you can choose the 2017 version. In this way, you can share your file with users who have an earlier version of *3ds Max*.

3ds Max allows you to save to formats up to 3 versions prior to the current version.

1.6.2 Save Selected

The *Save Selected* command allows you to save in a separate file only selected object(s). If, for instance, you click on the box and go *File > Save Selected*, rename it to box, and press *Save*, the box is saved in a new *3ds Max* file.

1.6.3 Archive

Archive is an important command, which allows you to move a file from one computer to another or exchange files. *3ds Max* works with paths. When you load a texture to apply it as a material (see Chapter 7) or an .ies file (see Chapter 5), *3ds Max* recognizes the path you follow to load these external files. If you change the location of any of these files or the location of the project, then your project will no longer render properly. *Archive* creates a compressed archive file that contains the scene file and any other files referenced by it, i.e., textures, ies, XRef, and so on.

To understand this option, open the file *Chapter 2.max*. Go to the menu *File* and choose *Archive*. The *File Archive* dialog box appears in which you define the destination folder. Give it a few minutes and a log window will appear (Figure 1.26).

Figure 1.26
Log window when archiving a project.

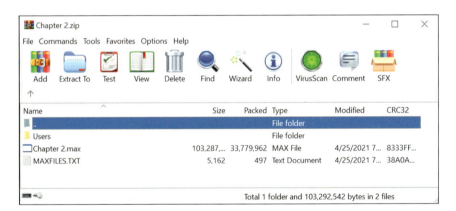

Figure 1.27
Files saved in the archive.

Once it disappears, the archive is created. Go to the destination folder you saved it. If you open it, you see the *.max* file along with folders that contain the maps used in this project (Figure 1.27).

> To share a project, first you need to archive it, and then share the zip file. If you only send the .max file, then all the textures and external files will be missing.

1.6 Saving a Project 17

2
Rendering Settings

The rendering settings are mainly algorithms that calculate effects not displayed in the viewports due to their complexity. Contrary to the viewport display, a render has lighting, shadows, and all the materials are displayed with their attributes, i.e., with reflections, refraction, bump effects, etc. V-Ray has many rollouts with commands to control the rendering settings. This chapter focuses only on the settings that are vital to control the quality of the renders relative to the rendering time. It is important to understand what each setting does, otherwise, you might end up doing test renders for several hours.

2.1 Assign V-Ray

Open the file *Chapter 2.max*. In this file, there is a camera placed, sunlight, artificial lighting, and materials, so it is the full final scene. To access the rendering settings, press *F10* or go to the menu *Rendering* and choose *Render Setup* or click on the *Render Setup* button, , from the *Main Toolbar* (Figure 2.1).

To use *V-Ray*, you must first select it as the current renderer. Thus, at the top of the *Render Setup* window, make sure the *Production Rendering Mode* is selected as the *Target* and then choose *V-Ray 5* as the *Renderer* (Figure 2.2).

If you only use *V-Ray* as a rendering engine, then you can set it as the default renderer to avoid repeating this step every time you start a new project. Thus, go to the menu *Customize* and choose *Custom UI and Defaults Switcher...*. Go to the *Initial settings for tool options* and choose *MAX.vray* (Figure 2.3). Click *Set* and restart *3ds Max* for the changes to take effect.

Figure 2.1

Part of the *Main Toolbar* and the *Render Setup* button.

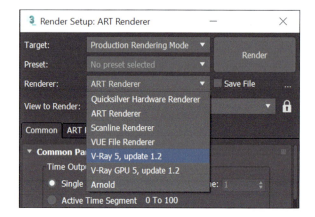

Figure 2.2

Selection of *V-Ray 5* as the renderer.

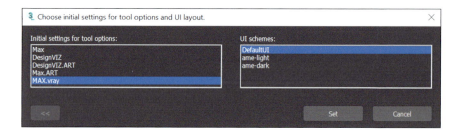

Figure 2.3

Set *V-Ray* as the default renderer.

2.2 V-Ray Frame Buffer

Before going into further details in the rendering settings, it is important to describe first the *V-Ray Frame Buffer* (*VFB*). The *VFB* is a window that displays the rendering progress and provides full control over the rendered output. In the *VFB* window, you can review, post process, and save a render.

2.2.1 Overview

To produce a render, activate the viewport you want to render, in our example, the viewport *VRayCam001*, and press *Shift+Q*. Alternatively, go to the menu

Figure 2.4

Part of the *Main Toolbar* and the *Render Production* button.

Rendering and choose *Render* or click on the *Render Production* button, ![icon], from the *Main Toolbar* (Figure 2.4).

When you execute the *Render* command, the active viewport will start rendering. So, make sure a camera viewport is active before you start rendering.

When the rendering process starts, three windows appear on the screen: the *VFB*, the *V-Ray Messages,* and the *Rendering* window (Figure 2.5).

If you click on the *VFB* window, it will get activated and will move to the foreground. The *V-Ray messages* and the *Rendering* window will hide behind it. In cases like this, move the *VFB* window around to locate the other two windows.

The render produced with the default settings is quite dark. Later in this chapter, we describe the steps to make a render brighter.

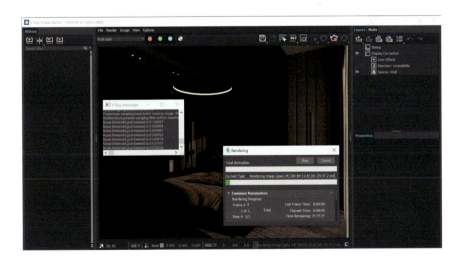

Figure 2.5

VFB, *V-Ray messages*, and *Rendering* windows.

2.2 V-Ray Frame Buffer

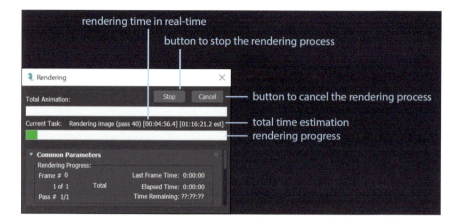

Figure 2.6

Rendering window.

2.2.1.1 Rendering Window

The *Rendering* window shows the progress of the render. More specifically, next to the *Current Task* there are two brackets, the first one shows the rendering time in real-time, while the second bracket estimates the total time needed for the render to be completed. Below the *Current Task* is a bar that displays in green color the progress of the render. Once the render is produced, the *Rendering* window disappears. There are two buttons on the *Rendering* window, *Stop* and *Cancel,* and they are both used to abort the rendering process (Figure 2.6).

The progress of the render is also displayed at the bottom of the *VFB* window (Figure 2.7).

Figure 2.7

Progress of the render displayed at the bottom of the *VFB.*

The main difference between *Stop* and *Cancel* is that if you use the *Bucket Image Sampler, Stop* will apply post-effects and save the result. In contrast, the *Cancel* button will end the rendering process without any post-render work done. However, if you use the *Progressive Image Sampler,* then regardless of whether you choose *Stop* or *Cancel,* post effects will be applied (see Section 2.3.1—*Image Sampler*).

2.2.1.2 V-Ray Messages Window

The *V-Ray Messages* window shows updates on the rendering process and if something is wrong, a "warning:" or "error:" is placed in front of the respective message.

2.2.1.3 VFB Window

The *VFB* window consists of the main view, where the render unfolds, and two side panels. The left panel is the *Render History*, where you can save, load, and compare renders. The right panel hosts the *Layers*, which allow you to do layer compositing and fine-tune the render directly in the *VFB* without the need of other software, like *Photoshop* (Figure 2.8).

Use the handles that divide each part to resize or collapse it. Drag the handles to resize the respective part or double-click on a handle to hide/appear that part (Figure 2.9). Rescale the *VFB* by clicking and dragging the edges of the *VFB* window. In addition, use the middle mouse button to alter the size of the image and to move the image within the *VFB* window.

2.2.2 Layers

The *Layers* allow you to adjust the rendered image by applying color corrections and other effects. A layer can be added either once the render has been processed or while the viewport is still rendering. Some of the most useful layers are:

Figure 2.8

VFB window—extended mode.

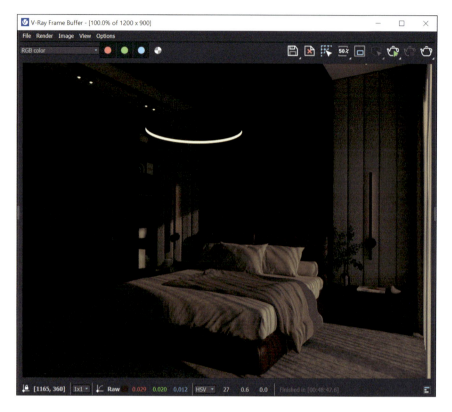

Figure 2.9

VFB window with the two side panels collapsed.

2.2.2.1 Exposure

The *Exposure* control includes three different settings, the *Exposure*, the *Highlight Burn*, and the *Contrast*. To enable the *Exposure* control click on the *Create layer* button, . From the drop-down list select *Exposure* and go to its *Properties* (Figure 2.10).

Exposure controls the brightness of the render. An *Exposure* value of +1.0 makes the image twice as bright compared to *Exposure* value 0.0, while −1.0 makes it twice as dark. In our example, set the *Exposure* value to 4 and the interior gets brighter. By increasing the exposure, you will notice that parts of the image render overexposured since the lights burn them (Figure 2.11).

Highlight Burn controls the highlights of the render, the parts that get burnt. The lower the *Highlight Burn* value, the less intense the bright areas of the render will be. Move the slider to the left and set it approximately to a value of 0.35. The areas will no longer look burnt.

Figure 2.10
Exposure layer.

Figure 2.11
Render with *Exposure*: 4.

2.2 V-Ray Frame Buffer

Figure 2.12

Rendered image with *Exposure*: 4.00, *Highlight Burn*: 0.35, and *Contrast*: 0.10.

Contrast is the ratio between the white and the black, or in other words, the light and the dark parts of a scene. In our example, set the *Contrast* to 0.10 to give more depth to the render (Figure 2.12).

In front of every layer, there is an eye icon, 👁. Click on it to disable and re-enable the respective layer (Figure 2.13).

2.2.2.2 White Balance

Click on the *Create layer* button and from the drop-down list select *White Balance* (Figure 2.14).

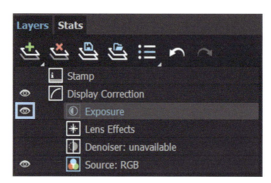

Figure 2.13

The eye icon in front of a layer disables/enables it.

26   2. Rendering Settings

Figure 2.14

White Balance layer.

White balance is a photographic term and refers to the process of removing any color cast or tint from a photo. Most light sources in a scene do not emit purely white color and have a certain color temperature. Incorrect white balance introduces a color tint, usually blue or yellow, to the renders and the result looks cool or warm. If all the lights you apply to the scene have a white color, then you probably will not need to adjust the *White Balance*. But if you use warm lighting, like 3500 K, then you need to adjust the *White Balance*, otherwise, it will look like there is an orange filter on the render.

To control the *White Balance,* go to the *Temperature* slider and slide to the left or right depending on the color tint you wish to remove from the image. In our example, you need to slide to the left to set the *Temperature* approximately to 5750 K (Figure 2.15).

To understand when to stop sliding, you need to have a white color as a reference. In our example, the ceiling and the bird accessory are white. When the temperature is set to 6500 K, which is the default value, they do not render white, but they have an orange hue. Start sliding to the left and at approximately 5750 K, when they turn into white, you can stop sliding. Be careful because if you slide a lot, they will start turning into a light blue color.

2.2.2.3 Curves

Curves control the contrast of the image. Click on the *Create layer* button, from the drop-down list select *Curves* and go to its *Properties* (Figure 2.16).

2.2 V-Ray Frame Buffer   27

Figure 2.15
Render with T*emperature*: 5750.

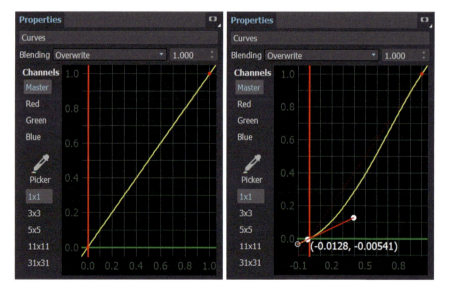

Figure 2.16
Plots of Curves without tangent lines (left) and with a tangent line at (0, 0) shifted downwards (right).

Figure 2.17

Final shape of curve.

At the plot of Curves, two red dots are visible—one placed at (0, 0) and the other one placed at (1, 1). Click on the bottom left red dot at (0, 0) and a tangent line will appear. Click and drag the tangent and move it slightly downwards and to the right (Figure 2.16—right). The contrast of the render increases, and the dark areas become darker. Then click on the upper right red point (1, 1) and another tangent appears. Click and drag the tangent and move it slightly upwards and to the left. The contrast of the render increases, and the bright areas become brighter. At the end you want the curve to form a smooth S, as seen in Figure 2.17.

2.2.2.4 Lens Effects

This layer is automatically added and is by default disabled. In the *Layers* tab, click on the *Lens Effects* and go to the *Properties*. Click in the *Enable bloom/glare effect* checkbox to enable it. A glare effect is automatically added to all the lights in the scene (Figure 2.18).

To control the glare effect, adjust the *Size* and the *Intensity* values (Figure 2.19).

Figures 2.20 and 2.21 show the render before and after adding the layers and enabling the *Lens Effects*. Note that in Figure 2.20, we have simply disabled the layers.

To delete a layer, select it and click on the *Delete Selected layers* button, .

Figure 2.18

Bloom/glare effect.

V-Ray keeps all the adjustments saved. Thus, if you close the *VFB* window and render again, the same layers will re-apply on the next render. To go back to the *V-Ray* default settings, go to the *Render Setup* dialog box *(F10)*, to the *Renderer* drop-down menu, pick another renderer, for example, the *ART Renderer* and then choose again *V-Ray 5* (Figure 2.22). The rendering settings will revert to the default ones.

Figure 2.19

Different glare effects depending on the *Size* and *Intensity* values.

Figure 2.20
Render before adding the layers.

Figure 2.21
Render after adding the layers.

2.2 V-Ray Frame Buffer 31

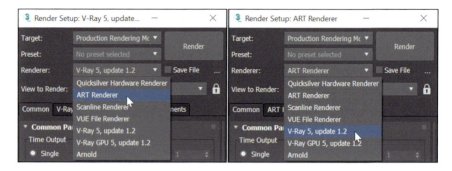

Figure 2.22

Switch between renderers to revert to the default *V-Ray 5* settings.

2.2.3 Saving a Render

You can save a render after it is cleared or while it is being rendered. In the *VFB* window, click on the *Save current channel* button from the *VFB toolbar* (Figure 2.23).

The *Save Image* dialog box appears. Type the desired name in the *File name* field, choose the file format in the *Save as type,* and choose the destination folder. Click on the *Save* button.

2.2.4 History

Render History allows you to save renders and compare them side by side in pairs of two or four. The render history images are saved as *V-Ray Image files* (.vrimg). This feature is useful especially when you are in the testing phase and you still try to figure out your solution.

To see how it works, first, produce a render. Go to the *History* panel, and you will notice that it is deactivated; you cannot press any of the buttons. Go to the *Options* menu of the *VFB* and choose *VFB settings* (Figure 2.24).

Go to the *History* tab of the *VFB settings* and make sure that *Enabled* is checked. In the *Location* field, if you enable *Use Project Path*, V-Ray will create a new *vfb_history* folder to use as the history folder within the current *3ds Max* project directory. If you disable *Use Project Path* and click on the *Save* button next to the path, you can specify the location where render history images will

Figure 2.23

Save current channel button.

Figure 2.24

Activation of the *History* panel through the *VFB settings*.

be saved. Enable *Use Project Path* and click *Save and close* for the changes to take effect (Figure 2.25).

The path in the *History* panel needs to be set only the first time. Every time you start a new project, *3ds Max* uses the latest path set in *History*.

History is now enabled. Click on the first button, *Save to history*, . A preview of the render will appear in the *History* panel. Leave the cursor on the preview and details of the produced render appear, like the resolution of the render, the time needed to be computed and the name of the *3ds Max* file (Figure 2.26).

We will now do one more test in which you will hide the fabric wall panels to see how the room looks like without them. If the render is still being calculated, press the *Stop* button on the *Rendering* window to stop the rendering process. Click on the wall panels to select them, right-click, and choose *Hide selection* (Figure 2.27). Produce a render.

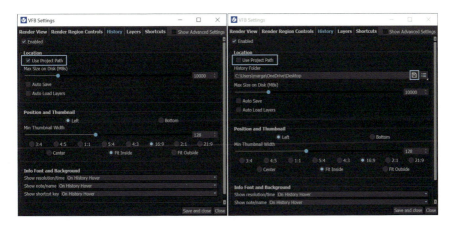

Figure 2.25

History save path.

2.2 V-Ray Frame Buffer

Figure 2.26

Preview of the saved render.

Notice in Figure 2.26—right that the render needed 48 minutes to be produced. You do not need to wait that long when you are in the testing phase. Once you have a fairly clean preview, press the *Stop* button on the *Rendering* window to stop the rendering process. Thus, in Figure 2.28, the render was stopped only after 5 minutes of rendering time. We do not have a clean render yet, but this render helps you to visualize the room and compare it to the previous option.

Click the *Save to history* button. Now there are two previews in the *History* panel. To compare them, click on the second button in the *History toolbar*, the

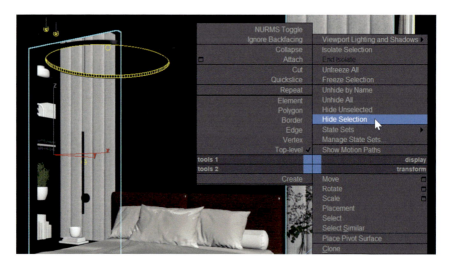

Figure 2.27

Hide selection command.

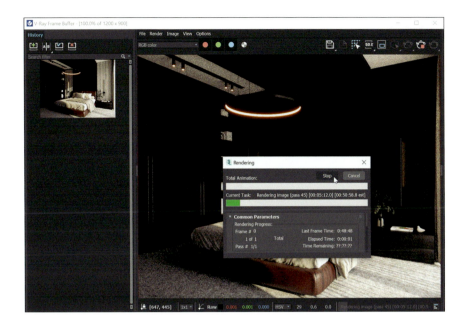

Figure 2.28

Stop button in the *Rendering* window.

A/B horizontal, ▣. Move the cursor to the right side of the first preview and when the *B* letter appears, click on the render. This way you set it as image *B* for the *A/B comparison*. Then move the cursor to the left side of the second preview and when you see the *A* letter, click on the render. You just set it as image *A* for the *A/B comparison*. If you see the render in the *VFB* window, the half left shows the old render and the half right shows the latest render. The two are divided in the middle by a white slider (Figure 2.29).

Click on the slider and move the cursor to the left or to the right. When you slide to the left, the B image unveils, while when you slide to the right, you see the A image (Figure 2.30).

To delete a render saved in the History panel, select the render and press the *Delete* button, ▣, from the *History toolbar*.

Note that to select a render in the *History* panel, you should first disable the *A/B horizontal* button, ▣.

Before moving on, unhide the fabric wall panels. Right-click anywhere in the viewport and choose *Unhide by Name*. From the *Unhide Objects* dialog box, select the wall panels and press *Unhide*.

2.2 V-Ray Frame Buffer

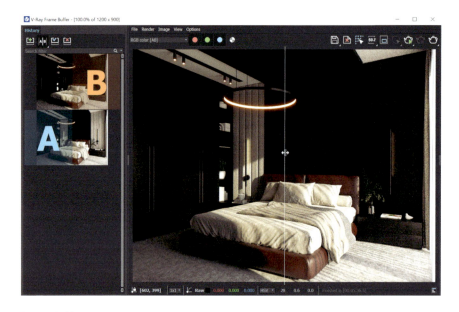

Figure 2.29
A/B comparison.

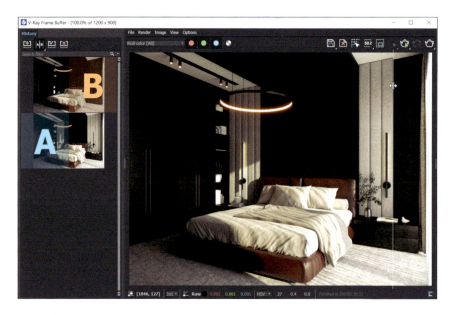

Figure 2.30
A/B comparison.

Now that you know how to use *VFB*, the next step is to start exploring the rendering settings.

2.3 Rendering Settings

Press *F10* to open the *Render Setup* window. The menus adjust depending on the selected renderer. When using *V-Ray 5*, the menus are the following: *Common*, *V-Ray, GI, Settings*, and *Render Elements*. The *Common* menu remains the same no matter which rendering engine is used. The *Common* tab contains commands that do not affect the photorealistic result and for this reason, we will explore this menu last. First, we will start with the *V-Ray* tab.

2.3.1 V-Ray

As mentioned at the beginning of this chapter, V-Ray has many rollouts and settings. This book focuses on the settings that are the most important ones to control the quality of the render. When rendering, there must be a fine balance between quality and time, and we are going to explore the settings that specifically address these two things.

2.3.1.1 Global Switches

Go to the *Global Switches* rollout and to the *Hidden lights* setting (Figure 2.31). By default, this option is enabled and it means that the lights will be rendered regardless of whether they are hidden or not. When this option is disabled, any lights that are hidden will not cast light in the rendering.

To understand how this option works, hide a light. More specifically, keep *Hidden lights* enabled, go to the *VRayCam001* viewport, select the right sconce, *Wall Sconce right*, right-click, and choose *Hide selection* (Figure 2.32).

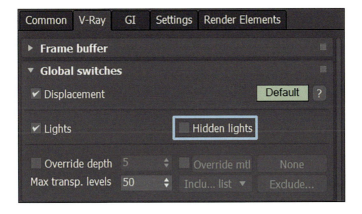

Figure 2.31

Hidden lights setting.

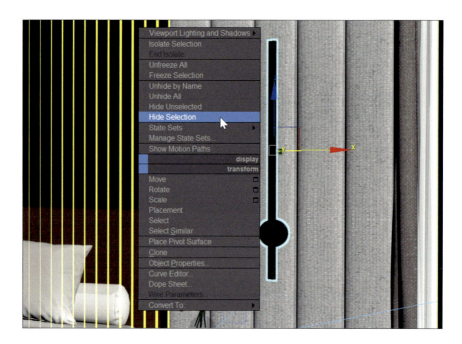

Figure 2.32

Hide Selection command using the right-click.

Press *Shift+Q* to produce a render. Although the wall sconce is hidden, it casts light on the wall panel (Figure 2.33).

If we go to the *Render Set* up dialog box, disable *Hidden lights*, and produce a render, the right wall sconce will no longer cast light (Figure 2.34).

Figure 2.33

Draft render with the right wall sconce casting light, although it is hidden.

Figure 2.34

Draft render with *Hidden Lights* disabled.

My advice is to always have the *Hidden lights* disabled. Especially if you are a beginner, you can get confused if you hide lights, and then in the render see illuminated areas, where no lights are visible.

Right-click anywhere in a viewport and choose *Unhide All* to unhide the wall sconce.

2.3.1.2 Image Sampler (Antialiasing)

An image sampler, as its name implies, is an algorithm that samples the image, which means, it calculates the color of the pixels. Each pixel in a rendering can have only one color and for *V-Ray* to determine the correct color, it samples colors from different parts of the pixel itself and around it. During the sampling process, the samples are added together and averaged to get the result. If you use a small number of samples, the result will be noisy. Noise in render is the grainy effect usually caused because the rendering engine uses a small number of samples.

In Figure 2.35—left, a small number of samples is used and, thus, noise is present in the render, while the right image has many samples and the render looks smooth and clean.

If you zoom in on the lumbar using the middle mouse button, you can see the pixels it consists of. During the sampling process, if you take the edge of the lumbar as an example, *V-Ray* gets one sample of the white pixel, one sample of the red pixel, and averages these two, which means that the pixel becomes a blend of white and red color (Figure 2.36).

You may now wonder, since you get a clean render only when you use many samples, why you do not always set many samples. The answer is because of the rendering time needed. It took only 5 minutes for Figure 2.35—left to render, while the right render needed 20 minutes. The rendering time is relevant to the

Figure 2.35

Render with a small number of samples (left) vs many samples (right).

computer specifications, and in your computer, the same render might need less or more time. But what you need to focus on here is the difference between the time needed when using many or few samples.

One of the purposes of sampling is also to improve antialiasing. To understand the meaning of antialiasing, consider the following example. When you

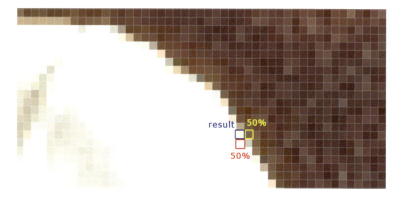

Figure 2.36

Sampling process: *V-Ray* samples the white color (red markup) and the red color (yellow markup) and puts an extra pixel of the average color of the two pixels (blue markup).

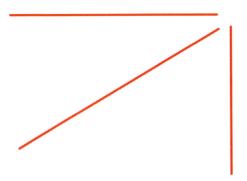

Figure 2.37

Antialiasing.

draw a straight line using any software, AutoCAD, Photoshop, 3ds Max, etc., and this line is horizontal or vertical, it appears as a perfect line. But if this line is at an angle, then you can see the pixels it consists of and the line is jagged (Figure 2.37). Antialiasing is a process for reducing the jagged distortions.

Going back to our example, antialiasing is basically the process of smoothing the lumbar to look like a smooth curve. To smooth the edges, *V-Ray* adds extra pixels.

If you use a small number of samples, the result will look grainy and pixelated, but it will render fast. While the more samples you add, the more accurate the result will be, but the longer the rendering time will be.

My advice is to use a few samples when you do your testing and once you are satisfied and ready to produce the final renders, then increase the samples to get a smooth result. Otherwise, the testing phase can take many hours.

V-Ray offers two methods for image sampling—the *Bucket* and the *Progressive*. They can be selected from the *Image sampler (Antialiasing)* rollout (Figure 2.38).

Figure 2.38

Types of image sampling.

2.3 Rendering Settings 41

The *Progressive* sampler is the default one and renders the entire image progressively in passes. In every pass, the render gets clearer. The *Bucket* sampler subdivides the image into rectangular sections, called buckets. The buckets render the image piece by piece. Regardless of the sampler you choose, they both do the same thing following a different procedure.

2.3.1.3 Progressive Image Sampler

In the *Image sampler (Antialiasing)* rollout select *Progressive* and go to the *Progressive Image sampler* rollout. The advantage of the *Progressive* sampler is that you can see a render very quickly and then let it refine for as long as necessary. The *Min. subdivs* and *Max. subdivs* control the minimum and maximum number of samples that each pixel in the image receives. The actual number of the samples is the square of the subdivisions. The default values of 1 and 100 produce a good number of samples. The *Render time (min-)* sets the rendering time in minutes. When this number of minutes is reached, the renderer stops no matter what the level of noise at that point is. If you set it to 1.0 and produce a render, the render will be calculated for 1 minute and then it will stop (Figure 2.39).

The noise in Figure 2.39 is high. The more you increase the time, the more passes are added in the rendering process, and so the cleaner the image becomes. Set the *Render time (min)* to 5.0 and produce a render. In Figure 2.40, the noise level is reduced, but it is still present in the render.

V-Ray has an extremely helpful feature called *VRayDenoiser*. It detects areas where noise is present in the render and smooths them out. To add *VRayDenoiser* go to *Render Setup*, select the *Render Elements* tab (last menu in the row *Common*,

Figure 2.39

Render with *Render time* set to 1 min.

Figure 2.40
Render with *Render time* set to 5 min.

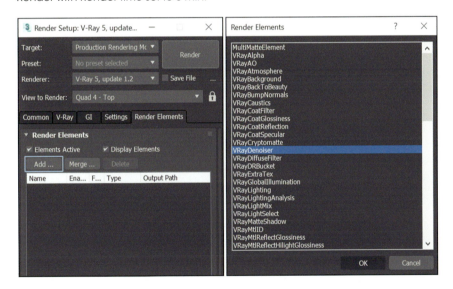

Figure 2.41
Steps to add *VRayDenoiser*.

V-Ray, GI, Settings, Render Elements), and click the *Add…* button. From the *Render Elements* window select *VRayDenoiser* and click *OK* (Figure 2.41).

Produce a *render* (Figure 2.42). Note that although the *Render Time* was kept to 5 minutes, the result is smooth due to the *VRayDenoiser*, which smartly blurs out the noise.

2.3 Rendering Settings

You should be careful when using *VRayDenoiser* because when blurring the image to eliminate the noise, it also blurs the textures and they lose their sharpness. My advice is for the final renders to always increase the samples so that the textures will render crisp.

If you set the *Render time* to 0.0, then you give preference to the *Noise threshold*. This setting controls the noise of the image. The value 0.005 is a good *Noise threshold*, it usually removes the noise and gives a clean image, but it can render for a long time. In our example, it took approximately 3.5 hours to render with *Noise threshold* at 0.005 and *Render time* set to 0 (Figure 2.43).

V-Ray gives priority to the first setting that will reach its standards. More specifically, if you set the *Render time* to 5 minutes and the *Noise threshold* to 0.005, the image will be rendering either for 5 minutes or until it reaches a noise level of 0.005 depending on which of the two settings is faster.

I would recommend that during the testing phase, where you will render multiple times, keep the rendering time low and add *VRayDenoiser* to speed up the process. Once you are ready for the final render, increase the rendering time.

If you set the *Rendering time* to 0 and the *Noise threshold* to 0.005, it will render for a long time. However, you do not have to wait until the render resolves completely. Press the *Stop* button to abort the rendering process once you are satisfied with the result (the noise level).

Figure 2.42

Render with *Render time* set to 5 min and *VRayDenoiser* added.

Figure 2.43

Render with *Render time*: 0 min and *Noise threshold*: 0.005.

2.3.1.4 Bucket Image Sampler

Go to the *Image sampler (Antialiasing)* rollout, select *Bucket,* and go to the *Bucket Image sampler* rollout (Figure 2.44). As we have already mentioned, the *Bucket* sampler subdivides the image into rectangular sections, called buckets, and the sampler renders the image piece by piece (Figure 2.45). The advantage of the *Bucket* sampler is that it requires less RAM, but the main disadvantage is that because it uses buckets, if you *Stop* the rendering process, it will only

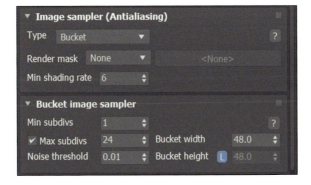

Figure 2.44

Bucket Image sampler.

2.3 Rendering Settings

Figure 2.45
The buckets while rendering.

clear the buckets that were calculated so far and not the whole image as with the *Progressive sampler*.

Two important parameters in this sampler are the *Min* and *Max subdivs*. To understand how they work, first disable the *Max subdivs*. Also disable *VRayDenoiser*. To do so, go to the *Render Elements* tab in the *Render Setup* dialog box, select *VrayDenoiser,* and press *Delete*. Disabling the *Max subdivs* basically means that you are disabling antialiasing and so all the objects will have staggered edges. Therefore, the render will be very noisy (Figure 2.46).

Enable back the *Max subdivs*. The default value is set to 24, which is a good number of samples to produce a render without noise, but it usually requires a substantial amount of rendering time. When you are in the testing phase, my suggestion is to use 4 samples in the *Max subdivs* to speed up the rendering process in combination with *VrayDenoiser*. When you are ready for your final render, you can make it 24.

Figures 2.47 to 2.49 show the same bedroom render using 4 *Max subdivs* without and with *VrayDenoiser,* and 24 *Max subdivs* with *VrayDenoiser*. The time needed in each case is also shown.

To sum up, with the image samplers, my advice is to use the *Progressive* method, because you can see a render very quickly and then let it refine for as long as necessary. For draft renders set the *Min subdivs* to 1, *Max subdivs* to 100, the *Render time* to 5 minutes, and *Noise threshold* to 0.01. Moreover, add *VrayDenoiser*.

Figure 2.46
Render with the *Max subdivs* disabled and *VRayDenoiser* deleted.

Figure 2.47
Render with *Max subdivs*: 4, without *VrayDenoiser*, *Render time:* 4' 43'', noise present.

2.3 Rendering Settings

Figure 2.48

Render with *Max subdivs*: 4, *VrayDenoiser* added, *Render time:* 4' 44'', no noise, blurred textures.

Figure 2.49

Render with *Max subdivs*: 24, *VrayDenoiser* added, *Render time:* 1 hour 07', no noise, crisp textures.

So, basically do 5-minute test renders. For final renders set *Min subdivs* to 1, *Max subdivs* to 100, the *Render time* to 0, and the *Noise threshold* to 0.005. You do not necessarily have to wait until the render resolves completely, because with these settings it might take several hours. Stop the rendering process when you see that the image is clean, and you are satisfied with the noise level.

If you are a beginner and find the *Progressive* method complicated on when to stop the render, then you can use the *Bucket sampler* and set for draft renders *Min subdivs* to 1 and *Max subdivs* to 4 and increase *Max subdivs* to *24* for the final renders.

2.3.2 GI

GI stands for *Global Illumination* and is a group of algorithms that are used to add more realistic lighting to a 3D scene. *Global illumination* is basically the indirect illumination, the illumination in a scene that comes from reflected (or bounced) light, as opposed to light coming directly from a light source.

2.3.2.1 Global Illumination

To understand how the indirect lighting works, go to the *GI* tab and open the *Global illumination* rollout. Uncheck the *Enable GI*, produce a render, and compare it to the render obtained with the GI enabled (Figure 2.50).

It is seen that when the *GI* is disabled, the scene gets darker since *V-Ray* calculates only the direct light of the scene, while when you enable GI, the light spreads in the room by bouncing on the surfaces. *V-Ray* offers several methods called *engines* for computing the *Indirect illumination*. The *Primary engine* is used to compute the first light bounce, and the *Secondary engine* is used to compute any subsequent bounces. As a *Primary engine* keep the default option, which is the *Brute Force*. It is the most accurate engine to render. The default

Figure 2.50
Render with the *GI* disabled (left) and enabled (right).

engine for the secondary bounces is the *Light Cache*, which offers an accurate result. Go to the *Light Cache* rollout. The *Subdivs* parameter controls the number of rays that are shot into the scene and the noise quality of the light cache samples. The default value of 1000 is a good number of rays to get a render without noise.

2.3.3 Render Elements

As mentioned earlier, one of the most important tools in the *Render Elements* menu is the *VRayDenoiser*. Use it to smooth out the render, especially when you use a small number of samples. Do not forget that *VRayDenoiser* will also blur the textures when you use a few samples. So, it is good for your draft renders, but always increase the number of samples for the final renders to have crisp images.

One more element that is very helpful is *VRayLightMix*. It allows you to quickly adjust the lighting of the scene by turning lights on and off, adjusting their intensity, and changing their colors, all from one control panel. See more information in Chapters 4 and 5. For now, click on the *Add* button and from the *Render Elements* list choose *VRayLightMix* and press *OK* (Figure 2.51).

> One thing you should keep in mind is that *VRayLightMix* works only with *Brute Force* and not with other engines. So always use *Brute Force* as the *Primary engine*.

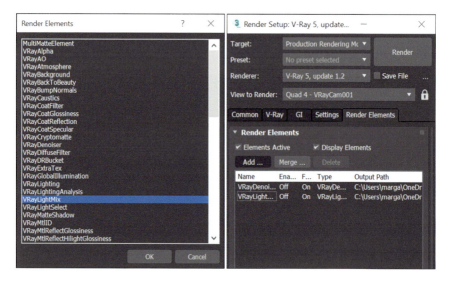

Figure 2.51

Addition of the *VRayLightMix*.

2.3.4 Common

In the *Render Setup* dialog box, the *Common* tab contains general controls for rendering, such as choosing to render a still image or an animation or setting the resolution of the rendered output. In the *Time Output* section, in the *Common Parameters* rollout, select *Single* to produce still images, renders. The other three options refer to animation, which this book does not cover. In the *Area to Render* section set which part of the viewport will be rendered. When *View* is selected, the entire viewport will be rendered (Figure 2.52).

With the *VRayCam001* viewport active, click on *View,* and from the drop-down menu select *Region*. A rectangular region appears in the active viewport. Click on the handles to adjust its size. When you hit the *Render* button, anything inside the region will be rendered, while anything that is outside this region will remain intact. As an example, adjust the size to include the accessories on the right nightstand (Figure 2.53).

If you render, the area outside the box remains untouched and *V-Ray* only renders the nightstand area (Figure 2.54).

When *Region* is enabled, you can only adjust the size of the rectangular region and move it to select a specific area of the scene. You cannot select specific objects from the scene. To unlock the *Region*, either click the *x* button at the top right corner of the region or right-click anywhere outside the region.

Even when you disable *Region, V-Ray* saves the selection. If you render, it will only render what is in the region, i.e., the accessories on the right nightstand. To render the whole scene again, you should select *View* from the *Area to Render*. After disabling the *Region* to re-enable it, go to the *Render Setup* window, to the *Area to Render*, and although *Region* is selected, click, and re-select *Region*. The handles of the region activate again.

Crop and *Region* are similar; they both render what is inside the region. The difference between them is that *Region* keeps intact anything outside the region, while *Crop* will only display what is in the region. For example, go to the *Area to*

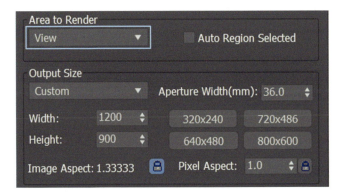

Figure 2.52
View selection to render the full viewport.

Figure 2.53

Region selection to render a specific area.

Render and choose *Crop*. If you render, only the part of the viewport that is inside the region is rendered and displayed (Figure 2.55).

Blowup and *Crop* do the same, i.e., they both render what is inside the region, without showing what is outside the region. The only difference is that when you use *Blowup,* the ratio of the region is fixed. Finally, there is the *Selected* option. Unfortunately, this option does not work with *V-Ray*, but only with the default *3ds Max* rendering engines.

The *Output Size* of the *Common* tab controls the proportions of the render that will be produced. You can do one of the following:

1. From the drop-down menu, select one of the real-world output sizes or choose *Custom* and type in the *Width* and *Height* fields the desired values.
2. Choose one of the template ratios on the right side, for example, 800×600.
3. Type the desired *Image Aspect,* say 1.333, if you want the width to be 1.333 times the height. Click on the *lock* icon next to it to lock this aspect and go to the *Width* and *Height* fields. Type the desired size in one of them and the other will be adjusted automatically. For our example, make the *Width* 1200 and the *Height* will become 900 (Figure 2.56).

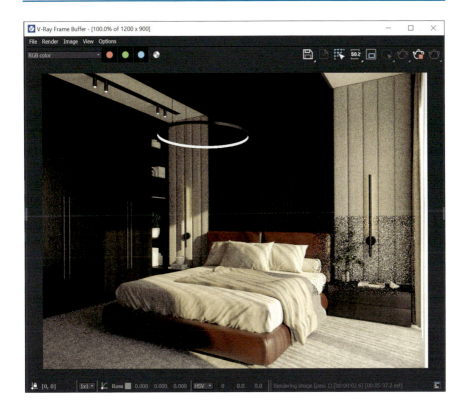

Figure 2.54
The area inside the region renders, while the area outside the region remains intact.

Figure 2.55
Crop render.

2.3 Rendering Settings   53

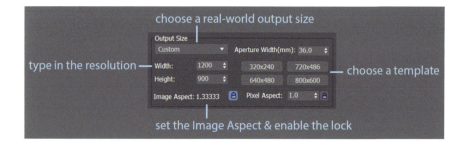

Figure 2.56
Output Size options.

If you adjust the aspect ratio and see the *VRayCam001* viewport, nothing changes. For the viewport to adjust to the new ratio, you need to enable *Show Safe Frames*. To do so, press *Shift+F* or click the camera viewport name label, *VRayCam001,* and select *Show Safe Frames* from the drop-down menu (Figure 2.57).

The viewport adjusts to the selected output size (Figure 2.58).

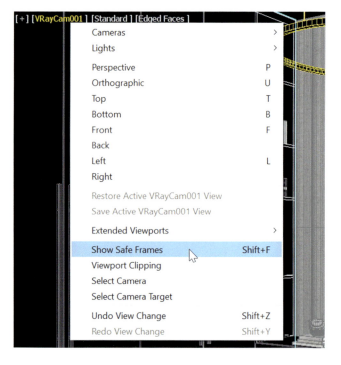

Figure 2.57
Show Safe Frames.

Figure 2.58

Show Safe Frames enabled.

To understand how the output size works, try a few more sizes to see how the viewport adjusts. If, for instance, you want to print the render in an A4 paper—horizontal orientation, go to the *Output Size*, click on the lock next to the *Image Aspect* to unlock it and type 3508 in the *Width* and 2480 in the *Height* fields (Figure 2.59).

To print the render in an A4 paper in the vertical orientation, go to the *Output Size* and revert the numbers, i.e., set the *Width* to 2480 and the *Height* to 3508 (Figure 2.60).

Figure 2.59

VRayCam001 viewport with *Output Size* 3508 × 2480.

2.3 Rendering Settings

Figure 2.60

VRayCam001 viewport with *Output Size* 2480 × 3508.

My suggestion is to set the aspect ratio from the very beginning of your project and always have the *Show Safe Frames* enabled. In any case, you can change the aspect ratio at any time during the work process. For our example, set the *Width* to 1200 pixels and the *Height* to 900 pixels. If you like the width to height ratio, it is good to click on the *lock* icon next to the *Image Aspect*. This way you can type a value in the *Width* field and the *Height* value automatically adjusts.

As a final step, scroll down to find the *Render Output* section to automatically save the render once it is computed. Click on the *Files* button to choose where you want the render to be saved and the file format (Figure 2.61).

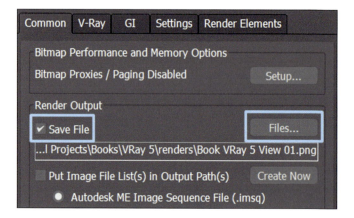

Figure 2.61

Save File option enabled.

56   2. Rendering Settings

When using the *Save File* option, be careful not to override the renders. Every time you produce a new render, you need to first click on *Files* to change the name of the new render. If you forget to change the name and press the *Render* button, *V-Ray* will warn you that the file already exists, and you will be asked to confirm to overwrite the image.

2.4 Interactive Production Rendering (IPR)

When you press *Shift+Q*, the rendering process begins. While you are rendering, you cannot make any changes to the scene. If you try, for instance, to select an object from the viewports, you will find that this is not possible. The rendering process must be completed first or paused. That holds unless you use the *IPR*. When you use the *IPR* method, you can make changes to the scene while the rendering is calculated, and the changes will be rendered in real-time. To start the interactive rendering, go to the *Render Setup* window (*F10*), to the *V-Ray* tab, and open the *IPR options* rollout. Click on the *Start IPR* button (Figure 2.62). Another way is to open the *VFB* window and click on the *Start interactive rendering* button on the *VFB* toolbar (Figure 2.63).

Figure 2.62

Start IPR button.

Figure 2.63

Start IPR button on *VFB* toolbar.

Figure 2.64

VFB window when using the *IPR* method.

While the scene is rendering, the *Start interactive rendering* icon changes to the *Refresh interactive rendering*. Click this button to restart the rendering at any time. The progress bar at the bottom of the *VFB* tracks the rendering progress. Use the *Abort rendering* button to stop *the IPR* (Figure 2.64). After stopping, you can click *Start interactive rendering* again to restart IPR.

To understand how *IPR* works, activate the *VRayCam001* viewport and start the interactive rendering. While the render is being calculated, click on the right wall sconce to select it. You will notice that you can select objects from the scene, while with the typical rendering process, this is not possible. With the sconce selected, right-click and choose *Hide selection* (Figure 2.65). The rendering automatically stops and restarts with the wall sconce hidden this time.

The interactive rendering allows you to save time, since you do not have to constantly start and stop the rendering process manually, but the program does it automatically. Moreover, it is helpful that you can select and edit items in the viewport while rendering.

Figure 2.65
Executing commands while rendering, when using the *IPR* method.

With the *IPR*, the rendering stops and starts from the beginning automatically every time you make a change.

The *IPR* method is mainly used while you are in the testing phase and you still work on the scene. My suggestion is to avoid it for your final renders since if you accidentally click on the screen, the render will start calculating from the beginning.

When you use the *IPR*, if you click in another viewport while rendering, for example, the *Top* viewport, it will automatically start rendering the *Top* viewport. To be able to make changes in all viewports and only render the camera viewport, go to the *Render Setup* dialog box, to the *View to Render* setting, select the camera, *VRayCam001*, and click the lock icon next to it (Figure 2.66). Now, regardless of the active viewport, only the camera viewport will render.

While you are rendering, you can also choose to do region renders. More specifically, while the *IPR* rendering is in progress, choose the *Region render* button and draw a region that contains the part you want to render (Figure 2.67).

You can click again the *Region render* button to disable it and render the full viewport.

The *Region render* works not only with the *Interactive Rendering*, but with the *Production* (typical) rendering as well.

2.4 Interactive Production Rendering (IPR)

Figure 2.66

Select the viewport you always want to render and click the lock icon.

Figure 2.67

Region render.

60 2. Rendering Settings

3
Cameras

Cameras present a scene from a particular point of view. The camera position impacts on the way we interpret a scene. A bird's-eye view gives the viewer an overview of the layout of the room. Low angles make the scene appear larger, while an eye-level view is a neutral position. Cameras in *3ds Max* and *V-Ray* follow the same rules as real-world cameras. They have settings like f-stop, film speed (ISO), shutter speed, and exposure value, which allow the simulation of a real camera and adjust the brightness of the image. In this chapter, you will learn how to place cameras and adjust their settings to achieve a photorealistic result.

3.1 Placing a VRayPhysicalCamera

Open the file *Chapter 3.max*. The file contains the bedroom scene you were working on in Chapter 2 with the only difference that in this file there are no cameras placed, since we are going to describe the steps to add them. To create a camera, go to the *Command Panel*, click on the *Create* tab and choose *Cameras*. Click on the *Standard* drop-down list, select *VRay*, and choose *VRayPhysicalCamera* from the *Object Type* rollout (Figure 3.1).

To place a camera, you need to click to set the position of the *Camera*, then drag and release the cursor to set the *Camera Target*. In our example, go to the *Top* viewport, click at the lower center of the room, drag the mouse, and release it at the upper center, as shown in Figure 3.2.

A *VRayPhysicalCamera* consists of two elements, the *Camera* and the *Camera Target*; and these two items are connected to each other with a line. When the *VRayPhysicalCamera* is selected, you can see the *Field of View*, which is the part of the scene that is visible by the camera. This part is defined by a triangle, as seen in Figure 3.3. Moreover, you can see the *Depth of Field (DoF)* or *Area of Focus*, which is the part of the scene that is on focus, i.e., the area where the objects look

DOI: 10.1201/9781003144786-3

Figure 3.1

Steps to select the *VrayPhysicalCamera*.

sharp in the render. This part is defined by the two outer lines inside the *Field of View* (Figure 3.3).

After placing the camera, the first thing you need to do is to change one of the viewports to show the camera's view. To do so, activate the *Perspective* viewport, click on the viewport name label, *Perspective*, and from the drop-down menu select *Cameras*. The *Cameras* submenu appears and shows the name of every camera in the scene. There is currently only one camera in the scene, so select *VRayCam001* (Figure 3.4).

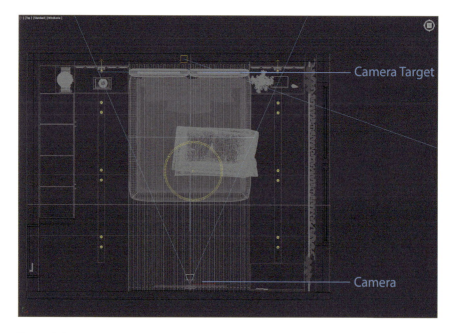

Figure 3.2

VrayPhysicalCamera placement.

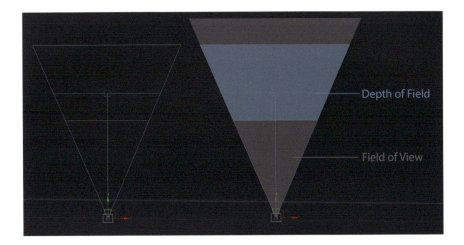

Figure 3.3
Representation of a camera showing the *Field of View* and the *Depth of Field*.

The viewport now shows the *VRayCam001* point of view. Moreover, you need to enable *Show Safe Frames (Shift+F)*. At a first glance, the point of view you get in our example looks odd, since one would expect to see the interior of the bedroom. The reason that the point of view looks like this is because anything you create in *3ds Max* is placed at height 0 cm. If you go to the *Left* view and check where the camera is positioned, you will see that it is basically on the floor. What you see now in the *VRayCam001* viewport is the bottom side of the bed and the *3ds Max* environment below that (Figures 3.5 and 3.6). Therefore, the next step is to select and elevate the *VRayPhysicalCamera*.

3.1.1 Selecting a Camera

When you create a camera in *3ds Max*, the camera is automatically selected. If it has been deselected, proceed to one of the following options to select it:

1. Press the *H* key. The *Select From Scene* dialog box appears, which shows in alphabetical order all the objects in the scene. Scroll down to find and select the Camera, *VRayCam001*, or the Camera Target, *VRayCam001*.

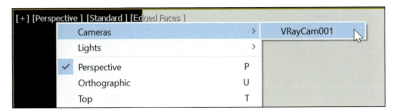

Figure 3.4
Steps to change the *Perspective* viewport to the *VRayCam001* viewport.

3.1 Placing a VRayPhysicalCamera

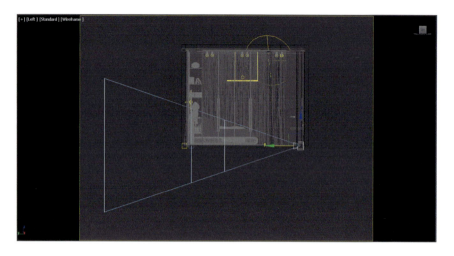

Figure 3.5

Default position of the camera at H: 0 cm.

Target. If you want to select both, choose one of them and then select the other one by holding down the *Ctrl* key. Then click the *OK* button (Figure 3.7—left).
2. Click on the camera viewport name label, *VRayCam001*, and from the drop-down menu choose *Select Camera* or *Select Camera Target* (Figure 3.7—right).
3. Click, directly from the viewports, on the Camera or on the Camera Target. If you click on the line that connects the Camera to the Camera Target, then both these two elements are selected.

Figure 3.6

VRayCam001 viewport.

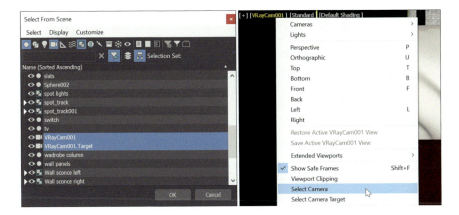

Figure 3.7

Select From Scene dialog box (left) and drop-down menu of the viewport *VRayCam001* (right).

Sometimes it is difficult to select a camera from the viewports because other objects are on top of it. An easy way to select specific types of objects is by using the *Selection Filter* from the *Main Toolbar*. The *Selection Filter* list lets you restrict the selection to specific categories. When the *Selection Filter* is *All*, you can select all the objects from the scene. However, if, for instance, you click on *All* and choose *Cameras*, then you can select only cameras from the viewports, whereas all the other objects in the scene freeze (Figure 3.8).

3.1.2 Using Transforms to Adjust the Camera Position

One way to move the camera in the scene to adjust its position is by using the transform commands and more specifically the *Select* and *Move* command. Therefore, either click on the *Select* and *Move* button from the *Main Toolbar* or press W. Then, go to the *Left* viewport and press the H key. Select both the Camera, *VRayCam001*,

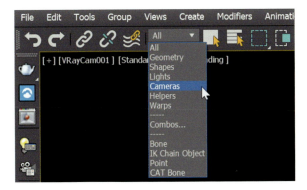

Figure 3.8

Use of *Selection Filter* to restrict the selection.

3.1 Placing a VRayPhysicalCamera 65

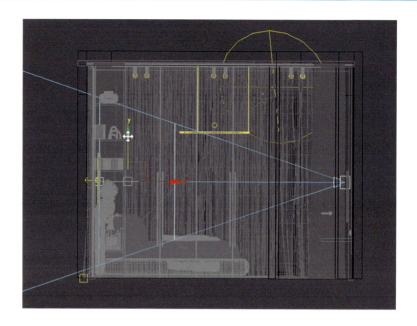

Figure 3.9

Elevation of the camera using the *Select and Move* command.

and the Camera Target, *VRayCam001.Target*, and click *OK*. Click on the *y* axis (green arrow) and drag the *VRayPhysicalCamera* higher (Figure 3.9). The *VRayCam001 viewport* updates as you are adjusting the position of the camera (Figure 3.10).

Figure 3.10

VRayCam001 viewport.

Figure 3.11
Setting the Camera and the Camera Target at 120 cm using the Z field of the Coordinates Display.

To place the camera at a specific height, for example, at 120 cm, you need to use the *Coordinates Display* in the *Status bar*. More specifically, select the Camera, go to the Z field of the *Coordinates Display* and type 120. Then select the Camera Target, go to the Z field and type also 120 (Figure 3.11).

Right-click in the camera viewport to activate the viewport without deselecting the camera. The camera viewport becomes active, but the camera is still selected in the other viewports. If you left-click in the camera viewport, you will lose your selection.

In the previous chapter, we set the *Output Size* to 1200 × 900.

If you see the camera viewport, only a zoomed part of the bedroom is visible, so you cannot really understand the bedroom design. The next step is to go to the camera settings to adjust them.

3.2 VRayPhysicalCamera Settings

If you go to the *Top* viewport to see where the camera is located, you will notice that it is placed right in front of the back wall. Although you have placed it at the very back of the room, you only see a zoomed part of it. Therefore, to capture more of the room, the first setting to adjust is the *Focal length* of the camera.

3.2.1 Focal Length

Select the Camera, *VRayCam001*, go to the *Modify* tab in the *Command Panel* to see the camera settings and go to the *Sensor & Lens* rollout. *Focal length* defines the width of the angle of view. The lower this value, the wider the angle. The default value is 40 mm. Type 30 and press *Enter* (Figure 3.12). If you check the camera viewport, you get the illusion that you stepped back although you did not move the Camera. The lens of the camera became wider.

Try 20 and press *Enter*. Now the camera captures all the bedroom, i.e., you can see all three sides of the room (apart from the back wall) (Figure 3.13).

My advice is to avoid using *Focal length* values lower than 25 to avoid any distortion of the scene. So, use 30 mm for our example and improve the overview of the bedroom by using the *Clipping planes*.

3.2.2 Clipping Planes

When a camera is selected, you see the field of view and that helps you to understand which part of the scene is visible in the camera viewport. In our example, based on the "triangle" of Figure 3.14—left, only a small part of the wardrobe and

Figure 3.12

VRayCam001 viewport with *Focal length*: 30 mm.

Figure 3.13
VRayCam001 viewport with *Focal length*: 20 mm.

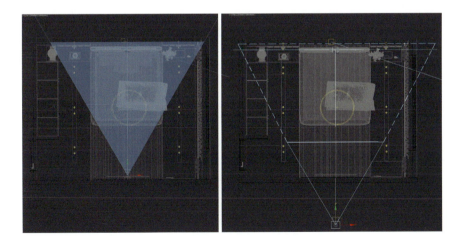

Figure 3.14
Current position of the Camera with the field of view highlighted with blue (left) and ideal position of the Camera to have a good overview of the room (right).

3.2 VRayPhysicalCamera Settings 69

the curtain is visible. So, the next step is to push the Camera further back and go behind the wall, to include a bigger portion of the closet and the curtain, as seen in Figure 3.14—right.

However, if you move the Camera behind the wall and look at the camera viewport, you only see the back wall of the bedroom. For this reason, the next step is to activate the *Clipping planes*, which remove this obstacle. With the *VRayCam001* selected, go to the *Modify* tab in the *Command Panel*. Scroll down, open the *Clipping & Environment* rollout, and enable *Clipping*.

To see the camera settings in the *Modify* tab, you need to select only the Camera, not the Camera Target or both.

Clipping planes let you to exclude some of the scene geometry and view only a certain portion of the scene. When enabled, two red planes come out of the camera. The one closer to the Camera represents the *Near clipping plane* and the other one is the *Far clipping plane*. Only the part of the scene that is between the near and the far planes is visible to the camera. Set the *Near Clipping Planes* to 165 cm or any value necessary, based on your Camera location, making sure that the near clipping plane is in front of the back wall (Figure 3.15). A red plane appears that shows where the distance of 165 cm is in relation to the location of the Camera. The back wall is now excluded from the camera and you can see the bedroom interior (Figure 3.16).

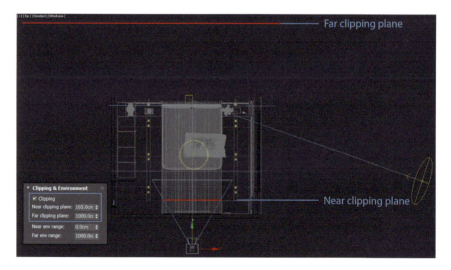

Figure 3.15

Near clipping plane: 165 cm (or any value necessary to be located in front of the back wall) and *Far clipping plane*: 1000 cm.

Figure 3.16
VRayCam001 viewport after adjusting the camera location and the *Near clipping plane* value.

If in the *Top* viewport you cannot see the red planes, zoom out. By default, the *Near clipping plane* is set to 0 cm, while the *Far* value is set to 1000 cm.

If you are using meters, the *Far* value is set to 1000 m and so you need to zoom out a lot to locate the far plane.

If you have the camera selected, in the camera viewport a grid of blue lines will also be visible. Deselect the camera for the blue grid to disappear.

When using the clipping planes, you must be careful where the camera is located. If the clipping planes intersect an object, they cut through that object creating a cutaway view and the cut part will render black (Figure 3.17).

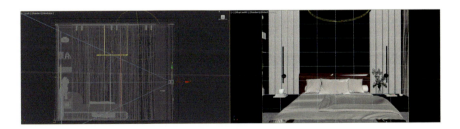

Figure 3.17
The *Near clipping plane* intersects the bed and cuts through it.

3.2 VRayPhysicalCamera Settings 71

3.2.3 Using the Camera Viewport Controls to Adjust the Camera Position

Another way to adjust the position of the camera is by using the *Camera Viewport Controls*.

At the right end of the *Status bar*, there are the buttons that control the display and navigation of the viewports. The *Navigation Controls* change depending on the active viewport. When you change an active viewport to a camera view, the *Navigation Controls* change to the *Camera Navigation controls*. As an exercise, we will use the *Camera Navigation Controls* to move the camera in front of the right nightstand.

With the *VRayCam001* viewport active, click and hold down the second command from the bottom row, *Walk Through*, 🚶, and select *Truck Camera*, 📷 (Figure 3.18). This command allows you to move both the Camera and the Camera Target parallel to the view plane. Move the camera to the left to align with the right nightstand (Figure 3.19—left).

When you use the navigation controls for a camera viewport, hold down the *Shift* key to constrain the movement to be vertical (or horizontal).

You do not need to have the Camera or the Camera Target selected to use the *Truck Camera* command. It acts on the active camera viewport.

Having completed the movement of the camera parallel to the view plane, use the first button from the upper row, the *Dolly Camera + Target*, 📷, to move the camera along its main axis, toward or away from what the camera is pointing at. If you hold down the *Dolly Camera* button, a flyout menu appears. You can select to either move the camera, *Dolly Camera*, or its target, *Dolly Target*, or both *Dolly Camera + Target*. Use the *Dolly Camera* command to move the camera closer to the nightstand (Figure 3.19—right).

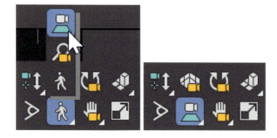

Figure 3.18

Hold down the *Walk Through* button and select the *Truck Camera* command.

Figure 3.19

Truck Camera and *Dolly Camera* commands used to adjust the position of the camera.

In this new position of the camera, the *Near clipping plane* intersects the bed and cuts through it. You do not need the clipping planes for this new position of the camera, so go to the camera settings and disable the *Clipping*. Repeat the commands *Truck Camera* and *Dolly Camera* until you get the result of Figure 3.20, i.e., a close-up view of the right nightstand accessories.

3.2.4 Simulation of Real-Life Cameras by VRayPhysicalCamera

As mentioned at the beginning of this chapter, a *VRayPhysicalCamera* simulates real-life cameras, so it is important to understand how a camera works. Figure 3.21 shows a simplified diagram of a camera.

Figure 3.20

VRayCam001 viewport after adjusting the position of the camera.

3.2 VRayPhysicalCamera Settings 73

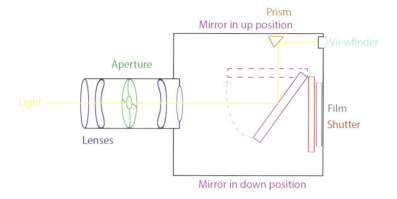

Figure 3.21

Simplified diagram of a camera.

In this very basic arrangement, a camera is essentially a box, with an opening at the front. In this opening, another box is attached and inside this box, there are multiple pieces of glass, the lenses. Light enters the camera through the lens and lands on the back of the camera. At the back of the camera there is the film and in front of the film there is the shutter, that basically protects the film. When you take a photo, the shutter will open just for a few seconds to exposure the film. Once the camera finishes taking the photo, the shutter closes. Before you click to take a photo, you need to see through the camera. For this reason, inside the camera, there is a mirror, placed in front of the shutter. When the mirror is in the down position, light comes in, it bounces off this mirror up into a prism ending into your eyes. So, when you click to take a photo, basically two things happen: the mirror lifts, the shutter opens, and the film gets exposed, i.e., captures the image. Then the shutter closes, the mirror returns to its original position, and you can see again through the lens.

Another issue closely related to a camera is the function of the human eye. During daytime when there is bright light, the iris constricts. In contrast, during nighttime, the pupil expands to allow more light to come in, for more visibility. Going back to the camera, between the lenses there is the *Aperture*, which basically resembles the human eye. It is spherical, with a hole to allow the light to come in and some blades that control the size of the hole and so the amount of light that enters the camera.

To take a good photo, you need to adjust the camera parameters. These parameters are the *Aperture*, the *Shutter speed*, and the *Film speed (ISO)*. The values of these parameters are basically combined into one value, the *Exposure value (EV)*. Figure 3.22 shows these four settings.

As already mentioned, *Aperture* is the opening, through which light enters the camera and falls to the sensor. The aperture is expressed in "f" numbers, where

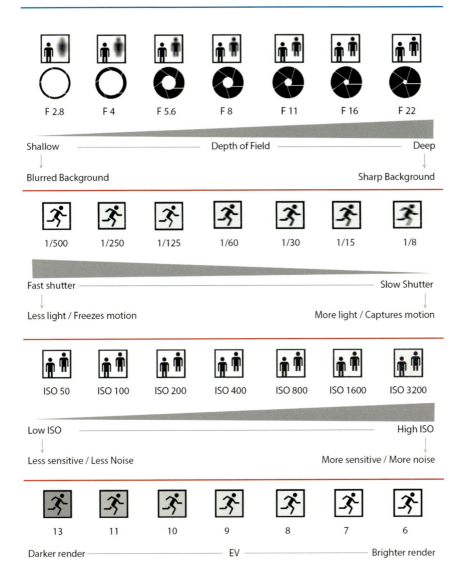

Figure 3.22

Aperture, Shutter speed, ISO, and EV diagram.

an f value is the ratio of the diameter of the lens aperture to the length of the lens. A typical range of f values is shown in Figure 3.22. The lower the f value, the larger the opening in the lens, the less *DoF*, resulting in blurred background. In contrast, the higher the f value, the smaller the opening in the lens, the greater the *DoF*, yielding sharper background. *Shutter speed* is the time the camera shutter is open. Therefore, it is the exposure time and for this reason, it is measured in

Figure 3.23
VRayCam001 render with the default settings.

fractions of a second, usually ranging from 1 s to 1/500 s. A slow shutter speed allows more light to hit the sensor, resulting in a brighter image. However, if there is motion, depending on the shutter speed, you may obtain freezing or blurring images since a fast shutter speed freezes the motion, whereas a slow shutter speed captures the motion and blurs the image. The *Film Speed* indicates the sensitivity of the sensor to light. It is expressed by the ISO number, which ranges from 20 to 3200. The higher the ISO, the more sensitive the sensor and the more light it collects. Low ISO means less sensitivity and the image is crisper. In night scenes, we usually need high ISO. The sensor becomes much more sensitive, but there will also be more noise. The combination of *Aperture*, *Shutter speed*, and *ISO* yields the *EV*. In general, the lower the *EV* the brighter the render.

We will now proceed to adjust these parameters into the virtual world in the *3ds Max* scene. Click on the *VRayCam001* viewport to activate it and produce a render. The result you obtain with the default settings is rather dark (Figure 3.23).

Select the *VRayCam001*, go to the *Modify* tab and to the *Color & Exposure* rollout. Click on the *Exposure* drop-down menu and select *Exposure value (EV)*. The default value is 13 (Figure 3.24). Noting that the lower you go the brighter the render, try 11 and produce another render. The scene is still a bit dark, so try an *EV* value equal to 8 (Figure 3.25).

However, the brighter the render, the more visible the highlights in the scene. Therefore, do not forget to apply the *Exposure* layer and adjust the *Highlight Burn* value (Figure 3.26).

Figure 3.24

Setting the *Exposure value*.

Figure 3.25

Render with *Exposure value*: 11 (left) and *Exposure value*: 8 (right).

Figure 3.26

Highlight Burn adjustment.

3.2 VRayPhysicalCamera Settings

Figure 3.27

Setting a *Physical exposure* (left) and the *Aperture* rollout (right).

Adjusting the *EV* is the easiest way to adjust the brightness of a render. But apart from the *EV*, you can also use the *F-Number, Shutter speed,* and *Film speed* to achieve the same result. More specifically, select the *VRayCam001*, go to the *Modify* tab and to the *Color & Exposure* rollout. Click on the *Exposure* drop-down menu, select *Physical Exposure*, and then go to the *Aperture* rollout (Figure 3.27).

Start the testing with the *Film speed (ISO)* setting, where the default value is 100. Based on the diagram of Figure 3.22 smaller values make the image darker, while larger values make it brighter. Instead of 100, type 1600 and render (Figure 3.28).

Figure 3.28

F-Number: 8, *Shutter speed*: 200, *ISO*: 100 (left), and *ISO*: 1600 (right).

Be careful that to adjust the settings in the *Aperture* rollout, you must select first *Physical exposure*. If you choose *EV*, then the *F-Number, Shutter speed,* and *Film speed* are disabled, so irrespective of whether you set *ISO* to 100 or to 1600, the result you obtain is the same.

Figure 3.29

F-Number: 8, *ISO*: 100, *Shutter speed*: 200 (left), and *Shutter speed*: 15 (right).

Higher *ISO* settings tell the camera sensor to group pixels together to capture more light. This grouping effect can make your image look noisy. To avoid that, instead of adjusting the *ISO*, you can play around with the *Shutter speed*. *Shutter speed* determines the exposure time for the virtual camera. The longer this time, the smaller the shutter speed values, the brighter the image. So, instead of 200, type 15 and produce a render (Figure 3.29).

Regarding the *F-Number* setting, wide aperture, i.e., low *F-Number* means more light falls on the sensor, but it also creates a short area of focus (*DoF*). Thus, instead of 8, type 1.2 and render (Figure 3.30).

As mentioned above, a low *F-Number* creates a short *DoF*, which means that parts of the render start to get out of focus and so they get blurred. But if you notice the latest render of *F-Number* 1.2, everything is crisp. This happens because *DoF* is by default disabled in *V-Ray*. When *DoF* is disabled, every object in the field of view looks sharp in the render irrespective of its distance from the Camera. To enable *DoF*, select the Camera, go to the *Modify* tab, to the *DoF* &

Figure 3.30

ISO: 100, *Shutter speed*: 200, *F-Number*: 8 (left), and *F-Number*: 1.2 (right).

3.2 VRayPhysicalCamera Settings

Figure 3.31

ISO: 100, *Shutter speed*: 200, *F-Number*: 1.2, and *DoF* enabled.

Motion blur rollout, and enable *DoF*. If you produce a render, you get Figure 3.31, where the accessories on the nightstand look all blurry.

To explain this result, you need to select the camera. In the *Top* viewport, you can see the characteristic triangle of the *Field of View* and the *DoF* defined by the

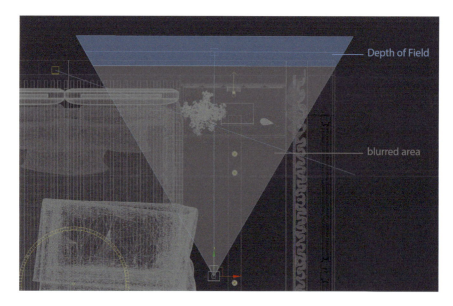

Figure 3.32

Depth of Field of the camera on the top view with *F-Number*: 1.2.

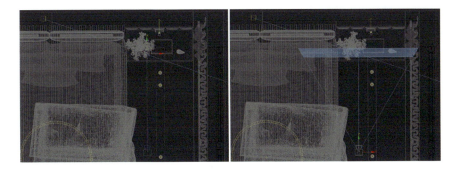

Figure 3.33
The position of the camera target affects the *Depth of Field*.

two outer lines (Figure 3.32). Note that objects in the center line of the *DoF* are captured very crisp, while objects at the edges and outside start to blur because they are out of focus. In our example, the *DoF* is beyond the nightstand and that is why the accessories on the nightstand look blurry in the render.

Select the Camera Target and move it to the front part of the nightstand. Then select the Camera and notice the new location of the area of focus (DoF) (Figure 3.33). If you produce a render, now the accessories will look crisp, whereas the back part gets gradually blurry (Figure 3.34).

When you increase the *F-Number*, the *DoF* increases as well. If you make it 8, the edges expand (Figure 3.35). So, if you now render, everything will be crisp, since

Figure 3.34
Render with the *DoF* adjusted.

3.2 VRayPhysicalCamera Settings

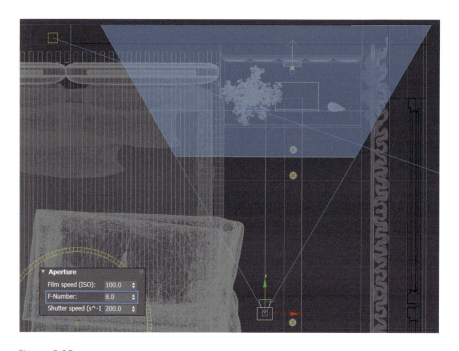

Figure 3.35

Depth of Field of the camera with *F-Number*: 8.

Figure 3.36

DoF enabled, *ISO*: 100, *F-Number*: 8, *Shutter speed*: 200 (left), and *Shutter speed*: 15 (right).

all the objects in the background are inside the *DoF*. However, do not forget that as you increase the *F-Number*, less light enters and, thus, the render becomes darker. So, when you adjust the *F-Number*, you also need to adjust the *ISO* or *Shutter speed* to control the brightness (Figure 3.36).

Figure 3.37

Part of the *V-Ray Toolbar—V-Ray Camera Lister* button.

Figure 3.38

V-Ray Camera Lister window.

If you choose *EV*, then the brightness of the render is adjusted only through the *EV* and the *F-Number* controls only the *DoF*.

When *DoF* is disabled, the render will be crisp, no matter the *F-Number* value.

3.3 Camera Lister

In *V-Ray 5, Update 1* a new feature is introduced, the *V-Ray Camera Lister*. It allows you to manage the settings of all the cameras in the scene from one panel. To access it, go to the *V-Ray Toolbar* and click on the *V-Ray Camera Lister* button, (Figure 3.37). The *V-Ray Camera Lister* window appears (Figure 3.38).

In this window, you can view the settings of all the cameras in the scene, you can change them, as well as copy and paste settings from one camera to another.

4

Natural Lighting

V-Ray offers two ways to set up the natural lighting of a scene—the *VRaySun* and the *Dome VRayLight*. Lighting in *V-Ray* simulates real-life conditions. This means that if you set the sunlight high in a scene, you get a strong, cold lighting with sharp shadows, while if you set it low, the lighting is warm with soft shadows. In this chapter, we describe how to use both these two methods, the *VRaySun* and the *Dome VRayLight*, as well as the use of the *VRayLightMix* to quickly navigate from one method to the other and create different lighting scenarios.

4.1 VRaySun

4.1.1 Placing VRaySun

Open the file Chapter 4.max. In this file, there is a camera placed with *Exposure value* set to 8. If you activate the camera viewport, *VRayCam001*, and produce a render, you will obtain a black image. This happens because at this point there are no lighting sources. To produce a render of a scene, you need to have at least one light source in this scene.

To create a *VRaySun*, go to the *Command Panel*, click on the *Create* tab, ➕, and choose *Lights* 💡. Click on the *Photometric* drop-down menu and select *V-Ray*. Click on *VRaySun* (Figure 4.1).

To place the *VRaySun*, right-click to set the position of the sun, drag the cursor, and release it to define the sun target. In our example, go to the *Top* viewport, right-click outside the bedroom on the right side, drag the cursor, and release it over the center of the headboard, as seen in Figure 4.2.

In the *VRaySun* pop-up window select *Yes*. In this way, V-Ray automatically places an environment in the scene, the *VRaySky*. *VRaySky* works together with *VRaySun* and they simulate real-life lighting. By changing the position of the sun,

Figure 4.1

Selection of *VRaySun*.

the sky also changes. As already mentioned, when you create something in *3ds Max*, it is placed at a height 0 cm. For this reason, if you go to the *Front* viewport, you see that the *VRaySun* is placed at the same height as the floor. Therefore, the next step is to elevate the sun.

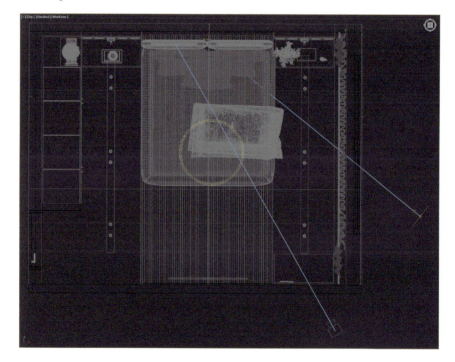

Figure 4.2

VRaySun placement.

4. Natural Lighting

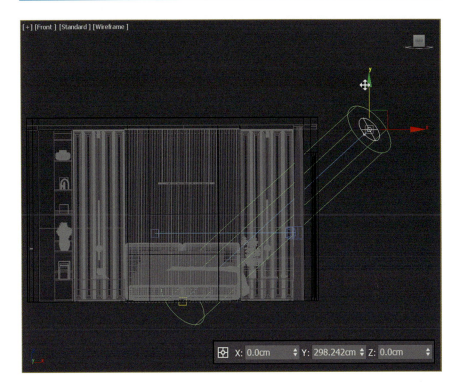

Figure 4.3
Repositioning of the *VRaySun*.

Enable the *Select* and *Move* command (*W*), go to the *Front* viewport, select the sun, and click on the *y* axis (green arrow). Drag it upwards at approximately 300 cm. When moving an object, you can check the coordinates fields in the *Status bar* or type there directly the desired value (Figure 4.3).

Now there is a light source in the scene, and you can produce a render. Since you are in the testing phase, do 5-minute test renders. Therefore, go to the *Render Setup* dialog box and set the *Render time* to 5 min, as described in Chapter 2, and produce a render (Figure 4.4).

The bedroom interior is now visible. The render is very noisy and the textures are blurry due to *VRayDenoiser*, but this should not trouble you at this point. For now, you are only interested in producing quick renders and once you are satisfied with the lighting, you will increase the *Render time* or set it to 0.0 and reduce the *Noise threshold* (0.005) to get a clear, crisp image.

If you forget to elevate the *VRaySun*, the render will be dark.

Figure 4.4

5-minute test render after placing the *VRaySun*.

4.1.2 VRaySun Parameters

As mentioned at the beginning of this chapter, *V-Ray* simulates how the sun works in real life. This means that the higher you place the sun the stronger and whiter the light will be. In contrast, the lower the sun, the softer and warmer the light. To apply these properties in our exercise, enable the *Select and Move* command, select the sun, go to the Z field at the coordinates of the *Status Bar* and type 600 cm. Produce a render. It is seen that, apart from the shadows that changed direction, the sun got stronger and whiter. Moreover, the shadows got sharper (Figure 4.5).

Now, move the sun position to Z = 80 cm and produce a new render. The bedroom gets darker, the sunlight is warmer and more subtle (Figure 4.6).

To increase the intensity of the sunlight, go to the *Sun Parameters* rollout and adjust the *Intensity multiplier*. Type 5 and render (Figure 4.7).

By increasing the intensity, the bright/overexposed areas of the render become also more intense. So, do not forget to always add the *Exposure* layer in the *V-Ray Frame Buffer* window and reduce the *Highlight Burn* value (Figure 4.8).

If you take a close look at the shadows, they are rather sharp, well defined. If you want them to look smoother, adjust the *Size multipier* in the *Sun parameters* rollout. When the *Size* is set to 1, the shadows are sharp. The higher you go, the smoother the shadows will be. Thus, type 7 and render (Figure 4.9).

Figure 4.5

Render with *VRaySun* position at Z: 600 cm.

Figure 4.6

Render with *VRaySun* position at Z: 80 cm.

4.1 VRaySun

Figure 4.7

Render with *Intensity multiplier*: 5.

Figure 4.8

Render with *Highlight Burn*: 0.30.

Figure 4.9

Render with *Size multiplier: 7*.

4.2 HDRIs and Dome VRayLight

An HDRI is basically a 360 photo (panorama) that contains a large amount of data, which can be used to emit light into a scene. You can find many HDRIs on the web. In *V-Ray 5, update 1* a new feature is added, the *Cosmos browser*, which includes a menu of free HDRIs that you can use in your scene. Thus, go to the *V-Ray Toolbar,* click on the *Cosmos browser* button, ▣ (Figure 4.10) and the *Chaos Cosmos Browser* opens.

There are three menus—the *3D Models,* the *HDRIs,* and the *Creators* menu, which is a combination of the first two. In this chapter, we will explore the *HDRIs* menu, while the *3D Models* menu is analyzed in Chapter 8. Click on the *HDRIs* menu and a list of free HDRIs appears organized into two categories, *Day* and *Night*. To use one of these HDRIs, you first need to download it. Choose *Day* and go to the *Day 035* preview. When you leave the cursor on a preview, the download button appears; either click on it to download the respective HDRI or click on the preview to open its parameters and then click on the *Download* button (Figure 4.11).

Once an HDRI is downloaded, the *Download* button changes to the *Import* button. You can either click on *Import* or simply drag and drop the *HDRI* in a viewport (Figure 4.12). It makes no difference where you place it. That is, whether it is placed at the top right side of the project or at the bottom left side, the result is the same.

Figure 4.10

Cosmos browser button.

Figure 4.11

Selection of the *Day 035* HDRI.

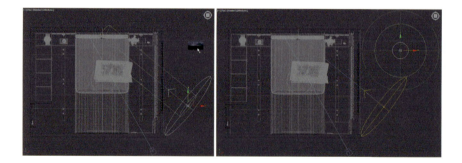

Figure 4.12

The drag and drop method to place an HDRI.

Download a few more HDRIs, for instance, the *Day 034* and the *Sunset 030*, and drop them in the *Top* viewport (Figure 4.13).

You now have three *Dome* lights and a *VRaySun* and all of them are enabled, which as a practice is not correct. In a scene, there must be one active *VRaySun* or one active *Dome VRayLight*. In Section 2.3.3, we added in the *Render Elements*

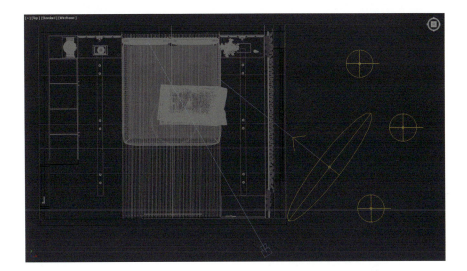

Figure 4.13

Multiple HDRIs placed in the scene.

tab of the *Render Setup* dialog box the *VRayLightMix* element (Figure 4.14). This element allows you to control all the lights of the scene from one panel.

Activate the camera viewport, *VRayCam001*, and click on the *Render* button to start rendering. Go to the *V-Ray Frame Buffer* window and to the *Layers* tab. Click on *Source: LightMix*, go to the *Properties* and choose *LightMix*. Note that there is a list of all the lights placed in the project (Figure 4.15).

You can click on *All* to disable all the lights and enable the first HDRI, the *Day 035 lightsource* (Figure 4.16). In this way, you can see how this HDRI lights up the scene without the distraction of the other light sources. Use the numeric field next to the light source to control the intensity of the respective light. If you type 2 instead of 1, you double the intensity of the *Dome VRayLight*.

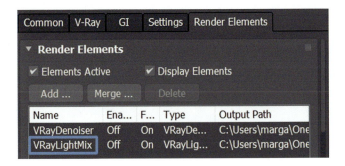

Figure 4.14

VRayLightMix added as a render element.

4.2 HDRIs and Dome VRayLight

Figure 4.15
LightMix showing a list of all the lights of the scene.

Figure 4.16
Render with only the *Day 035* HDRI enabled.

Figure 4.17

Render with the *Day 034* HDRI enabled (left) and with the *Sunset 030* HDRI enabled (right).

Then you can disable *Day 035* and enable *Day 034*. Repeat the steps for the *Sunset 030* HDRI (Figure 4.17).

So, placing multiple HDRIs in combination with *VRayLightMix* allows you to do multiple tests quickly on the lighting of the scene to achieve the aesthetic result you want.

4.2.1 Dome VRayLight Settings

If you look at the *Dome VRayLight*, there is a line, a pointer at its right side (Figure 4.18). This line indicates the center of the HDRI image. Usually, the sun is placed at the center of the HDRI image, but always check the HDRI preview to confirm that (Figure 4.19).

If you want to change the direction of the shadows, you need to rotate the *Dome VRayLight*. Select the *Day 034* Dome light and rotate it by 50 degrees. For the rotation, use the *Select and Rotate* command from the *Main Toolbar*, click on the *Dome VRayLight*, and type −50 in the Z field in the *Transform Type-In* fields. Before rendering you need to go to the *Modify* tab, to the *Dome light* rollout,

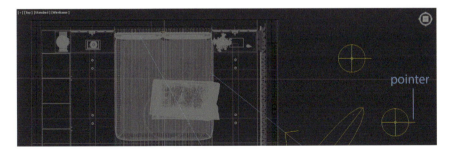

Figure 4.18

The pointer on the *Dome VRayLight*.

4.2 HDRIs and Dome VRayLight 95

Figure 4.19

The blue line indicates the position of the sun and the center of the HDRI.

and enable *Lock texture to icon* (Figures 4.20 and 4.21). If you do not enable this option, then no matter how you rotate the *Dome* light, the sun remains in its original position.

If you want to increase the intensity of the *Dome* light, you can use the *LightMix* panel, as described above. An alternative way is to select it, go to its properties, and increase the *Multiplier*. If instead of 1 you type 3, you will get the render seen in Figure 4.22.

If you are using a version of *V-Ray* prior to 5, then you will not have access to the *Cosmos browser*. In this case, you need to create the *Dome VRayLight* from the *Command Panel*. More specifically, go to the *Command Panel*, to the *Create* tab, choose *Lights,* and from the *V-Ray* menu select *VRayLight*. Go to the *General* rollout and from the *Type* drop-down menu click on *Plane* and choose *Dome* (Figure 4.23).

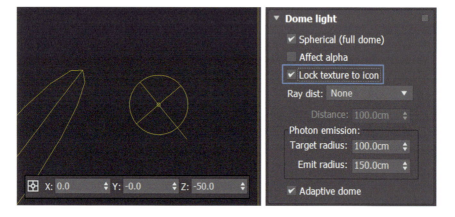

Figure 4.20

Day 034 HDRI rotation by 50° (left) and the *Lock texture to icon* setting enabled (right).

Figure 4.21
Render after rotating the *Dome* light by 50°.

Figure 4.22
Render with the *Dome VRayLight Multiplier*: 3.

4.2 HDRIs and Dome VRayLight   97

Figure 4.23

Selection on the *Dome* type.

Go to the *Top* viewport and click to place the *Dome VRayLight*. To load an HDRI, go to the *Dome VRayLight* parameters and to the *General* rollout click on the *No Map* button and choose *Maps > VRay > VrayBitmap* (Figure 4.24). Choose an HDRI from the web and click *Open*.

> You can find many HDRIs on the web; a website I could suggest that offers free HDRIs is hdrihaven.com.
> If you have previous versions of *V-Ray*, then you need to choose *VRayHDRI*. If you have *V-Ray 5*, choose *VRayBitmap*.
> If you do not add an HDRI, you basically set a diffuse light in the scene. The objects cast shadows, but you do not have directional light and shadows from the sun.

To sum up, there are two options to light a scene naturally, either using the *VRaySun* or the *Dome VRayLight*. When using the *Dome VRayLight*, you can either use it in its default form and light the scene uniformly or add an *HDRI* to have directional light and shadows.

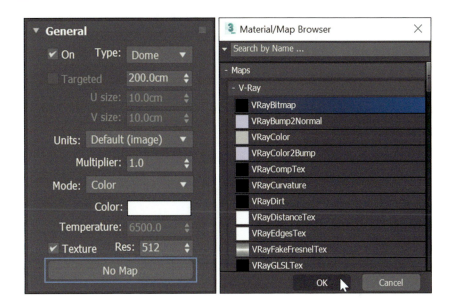

Figure 4.24
Selection of *VRayBitmap* using the *No Map* button.

4.2 HDRIs and Dome VRayLight 99

5
Artificial Lighting

When we work on interior scenes, apart from the natural lighting, the artificial lighting also plays an important role in the composition and final result. It is used to create different scenarios of the same scene, from a daytime view to a nighttime shot. *V-Ray* includes a set of lights designed specifically for rendering with the *V-Ray* engine. We have seen so far the *VRaySun* and the *Dome VRayLight*. In this chapter, we explore the *VRayIES* lights and more options in the *VRayLights*. Moreover, we describe the use of the *VRayLightMix* to control the lighting of the scene.

5.1 VRayLights

Open the file *Chapter 5.max*. In this file, there is a *VRayPhysicalCamera* of *Exposure value* 8 and a *VRaySun*. The *Output size* is set to 1000 × 850. If you produce a render, you obtain Figure 5.1. The only light source in the scene is the sunlight, a *VRaySun*. The next step is to place artificial lights, i.e., the lighting that emanates from electric lamps. As in the previous chapter, set to produce 5-minute test renders since at this point it is more important to do quick renders, and once the result is satisfying, change the settings to produce the final renders.

> Do not forget to apply the *Exposure* layer to adjust the *Exposure*, the *Highlight Burn,* and the *Contrast* of the render. In our example, set the *Highlight Burn* to 0.30 and the *Contrast* to 0.10.

The *VRayLight* is a *V-Ray* light source that is used to create physically accurate lights of different shapes, like plane, sphere, etc. The first *VRayLight* we will place is the one behind the two wall fixtures. In general, when you want to place a light on one or more objects, it is easier to isolate these objects, so that you will not get

DOI: 10.1201/9781003144786-5

Figure 5.1

Draft render of the *VRayCam001* viewport.

confused by the projection lines of the other objects in the scene. Therefore, select the two wall fixtures, right-click and choose *Isolate Selection* (Figure 5.2). In the viewports only the two wall sconces are visible.

Go to the *Command Panel*, click on the *Create* tab, and choose *Lights*. Click on the *Photometric* drop-down list and select *VRay*. Choose *VRayLight* from the *Object Type* rollout (Figure 5.3).

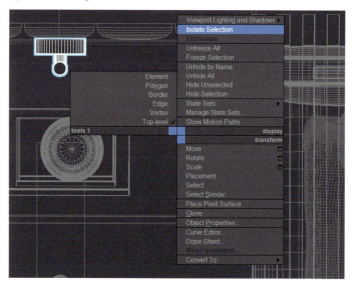

Figure 5.2

Selection of the two wall sconces and activation of the *Isolate Selection* command.

Figure 5.3

Selection of *VRayLight* from the *VRay* category.

Go to the *Front* viewport, click on the top-left corner of the right sconce and drag diagonally to the bottom right corner of the sconce (Figure 5.4—left). It does not matter if the *VRayLight* fits exactly the dimensions of the light fixture since you will adjust its size in the *Modify* tab. Enable the *Select* and *Move* command, go to the *Top* viewport, press the *Zoom Extends All Selected* command,

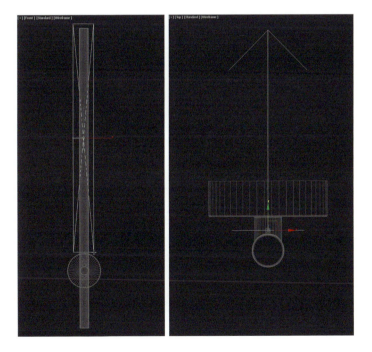

Figure 5.4

Creation of *VRayLight* in the *Front* viewport (left) and its relocation in the *Top* viewport (right).

5.1 VRayLights

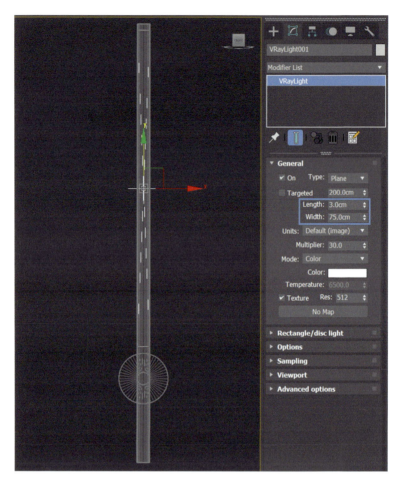

Figure 5.5
Control the size of the *VRayLight* in the *Modify* tab.

from the *Navigation Controls*, find the *VRayLight,* and drag it to the correct location, as seen in Figure 5.4—right.

With the light selected, go to the *Modify* tab in the *Command Panel*. Go to the *General* rollout and resize the *VRayLight* by adjusting the *Length* and *Width* values. Make the *Length* 3 cm and the *Width* 75 cm (Figure 5.5).

To make a copy of this light for the left wall fixture, hold down the *Shift* key and drag the light. When you release the cursor, the *Clone Options* dialog box appears. Select *Instance* and click *OK* (Figure 5.6).

By choosing *Instance*, an interchangeable clone of the original *VRayLight,* the *VRayLight002,* is created. This means that if you modify one of the two *VRayLights,* the other one is modified simultaneously. If you choose *Copy,* a

Figure 5.6

Creation of an *Instance* of the *VRayLight*.

separate clone from the original item is created; modifying one *VRayLight* has no effect on the other. If necessary, move the *VRayLight002* properly behind the left wall fixture. Right-click in any viewport and select *End Isolate* to end the isolation mode and return to the full scene.

When you are doing test renders, it is useful, instead of rendering the full viewport, to use the *Region* or *Crop* method to render only the part of the scene you are working on. Press *F10* to open the *Render Setup* dialog box, go to the *Common* tab, and the *Common Parameters* rollout. Go to the *Area to Render* section, click on *View*, and select *Region*. Adjust the region box to include the area around the two wall sconces, as shown in Figure 5.7, and produce a render (Figure 5.8). Note that the *Region* renders only the part of the viewport that is in the region box and leaves the outside part as it was in the previous render.

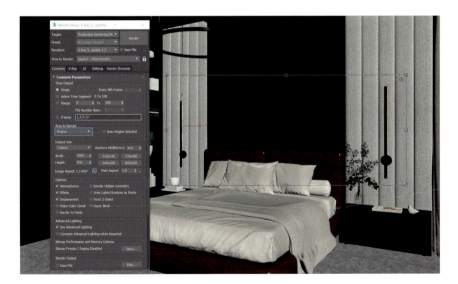

Figure 5.7

Adjustment of the region to include the two wall sconces.

5.1 VRayLights

Figure 5.8
Region render of *VRayCam001*.

With one of the *VRayLights* selected, since they are instances (see Figure 5.6), go to the *Modify* tab and to the *Mode* in the *General* rollout. This parameter specifies the mode in which the color of the light is determined. When the *Color* option is enabled, the color of the light is specified by the color swatch, while if you choose *Temperature*, the color of the light is expressed in Kelvin. Select *Temperature*, go to the *Temperature* field, and type 3500. Produce a render (Figure 5.9).

The light intensity can be measured in 5 different ways. Go to the *Units* in the *General* tab to select the desired method. If you choose *Luminous power (lm)* or *Radiant power (W)*, the intensity is measured in lumens and Watts respectively and the intensity of the light does not depend on its size. If you choose the other 3 types, *Default (image)*, *Luminance,* or *Radiance*, then the size of the *VRayLight* as well as the *Multiplier* value affects the intensity. For our example, keep the *Default* units. If you want to turn off a light, simply disable the *On* checkbox in the *General* rollout.

When you place a *VRayLight*, make sure that it is placed in front of the object and not in the object. If you accidentally place it in the object, for example, in the sconce, then it will not cast light. So, always zoom in to confirm the correct position of the light.

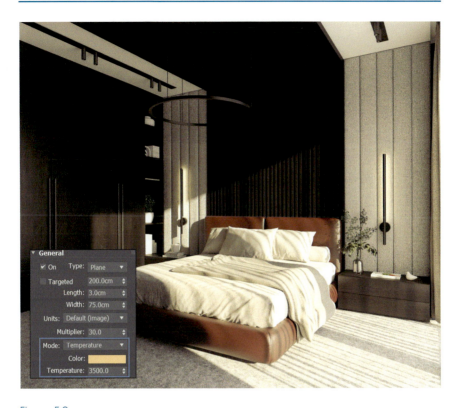

Figure 5.9

VRayLight Temperature set to 3500.

The *VRayLight* has an arrow on one side. This arrow indicates the direction of the light. If you want the *VRayLight* to cast light on both sides, then go to the *Modify* tab and enable *Double-sided* from the *Options* rollout (Figure 5.10).

If you zoom in the left wall sconce, you will notice that the shape of the *VRayLight002* is partially visible on the right side of the sconce and it renders as a black shape. To understand this better, hide the wall scone and produce another render. The *VRayLight* renders as a bright white plane from the side of the arrow and as a black plane from the opposite side (Figure 5.11). If you do not want the shape of the *VRayLight* to be visible, select it, go to its properties in the *Modify* tab and enable *Invisible* in the *Options* rollout (Figure 5.12).

To control the intensity of the *VRayLight* either go to the *Modify* tab and to the *Multiplier* value or use the *LightMix* panel.

The next step is to place the lights on the ceiling track. Select the two tracks (*spot_track* and *spot_track001* in the *Select by Name* list), right-click and choose *Isolate Selection*. Go to the *Command Panel*, click on the *Create* tab and select *VRayLight*. Go to the *General* rollout, click on the *Type*, and select *Sphere*. Right-click in the *Top* viewport to activate it and zoom in on one of the spotlights.

5.1 VRayLights

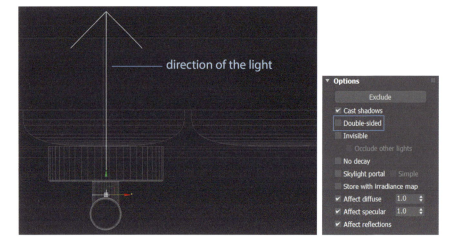

Figure 5.10

The arrow on the *VRayLight* indicates the direction of the light. When *Double-sided* is enabled, it casts light on both sides.

Figure 5.11

The shape of the *VRayLight* is visible and renders black (left). Sconce is hidden to see the shape of the *VRayLight* (right).

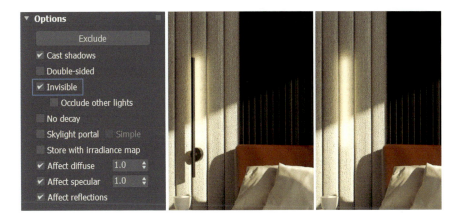

Figure 5.12

If *Invisible* is enabled, the shape of the *VRayLight* does not render.

Click and drag to create a spherical *VRayLight* (Figure 5.13). Right-click in the *Front* viewport, activate the *Select* and *Move* command, zoom out to find the *VRayLight*, and place it in the correct position in the spotlight (Figure 5.14).

Go to the *Top* viewport, hold down the *Shift* key to make an instance of this spherical *VRayLight*, and drag it in the spotlight next to it (Figure 5.15).

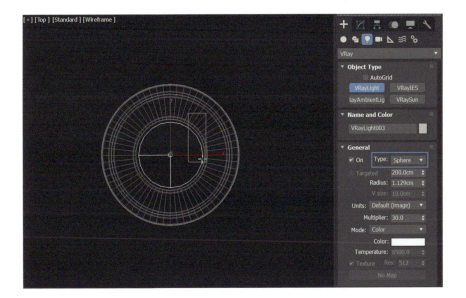

Figure 5.13

Creation of a *VRayLight* using the *Type: Sphere*.

5.1 VRayLights

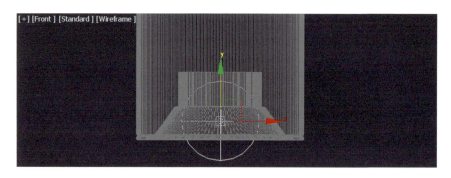

Figure 5.14
Repositioning of the *VRayLight* in the *Front* viewport.

Figure 5.15
Creation of an instance for the second spotlight.

Go to the *Selection Filter* in the *Main Toolbar* and instead of *All*, choose *Lights*. Use the *Select* and *Move* command and make a window to select both *VRayLights* (Figure 5.16). Zoom out to see the other spotlights, hold down the *Shift* key, and make instances for the pair of the spotlights in the middle of the track. Repeat the steps to put *VRayLights* in all the spotlights. Do not forget to make instances for the right ceiling track lights as well.

Right-click and choose *End Isolate*. Go to the *Render Setup* dialog box and make a *Crop* render at the ceiling track lights (Figure 5.17).

Select one of the *VRayLights*, go to the *Modify* tab, select *Temperature*, make it 3500, and produce a new render. Do not forget to select the *Lens Effects* in the *V-Ray Frame Buffer* window and *Enable bloom/glare effect* to add a glare to the

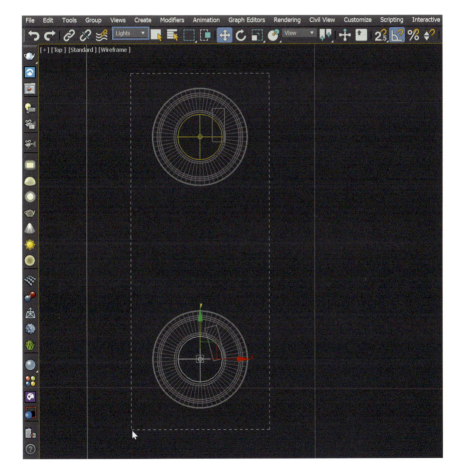

Figure 5.16

Window selection of both *VRayLights*.

5.1 VRayLights

Figure 5.17

Crop render of the track lights.

lights (Figure 5.18). To control the glare effect, adjust the *Size* and the *Intensity* values.

To see the glare of a light, you need, in the *VRayLight* properties, to have the option *Invisible* disabled. If you enable *Invisible*, the *VRayLight* casts light, but the shape of the light is no longer visible.

The next light we will place is the light of the pendant. So far, you have only placed *VRayLights* that have a rectangular or a spherical shape, but this pendant is circular, so you can use neither the *Plane VRayLight* nor the *Sphere VRayLight*. To create lights of random shapes, use the *Mesh* type in the *VRayLight* parameters. Thus, select the pendant, right-click and choose *Isolate Selection*. If you cannot select the pendant from the viewports, do not forget that you might still have the *Lights Selection Filter*. Go to the *Selection Filter* and choose *All*. Otherwise, use the *Select by Name* command (*H*).

Figure 5.18

Bloom/glare effect enabled.

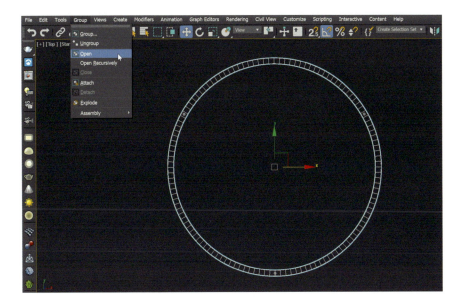

Figure 5.19

Isolation of the pendant to open its group.

The pendant is a group. To open a group and see the parts it consists of, select the pendant, go to the menu *Group* and choose *Open* (Figure 5.19).

When a group is open, you can either click on the object you want to select or press the *H* key. The *Select From Scene* dialog box appears. Go to the *pendant* and click on the arrow next to its name to see the parts it consists of. The pendant consists of two parts—the *light* and the *metal* part. Select the *light* and click *OK* (Figure 5.20).

You can understand when an object is a group, because when you select it and check the *Name and Color* field in the *Command Panel*, the name appears with bold letters.

The object *light* is the part of the pendant that needs to be converted to a light. Right-click and choose *Isolate Selection* so that the remaining pendant will not confuse you. Go to the *Command Panel*, click on the *Create* tab, and select *VRayLight* from the *Lights* and the *VRay* category (see Figure 5.3). Randomly draw a *VRayLight* as seen in Figure 5.21.

With the *VRayLight* selected go to the *Modify* tab, to the *General* rollout, and to the *Type* setting. Click on *Plane* and choose *Mesh*. The *VRayLight* automatically gets a boxed shape. Go to the *Mesh light* rollout and click on the *Pick mesh* button (Figure 5.22). Click on the object *[pendant] light*. The *VRayLight* automatically converts to the shape you clicked on and you automatically return to the full scene (Figure 5.23).

5.1 VRayLights   113

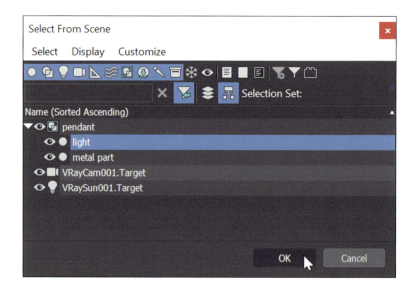

Figure 5.20
Selection of the *light* of the pendant group.

Go to the *Modify* tab, select *Temperature,* and type 4000. Produce a cropped render of the pendant (Figure 5.24). So, when you use the *Mesh* type, the *VRayLight* can get any shape, if you first have modeled the shape/object and then you just click on it.

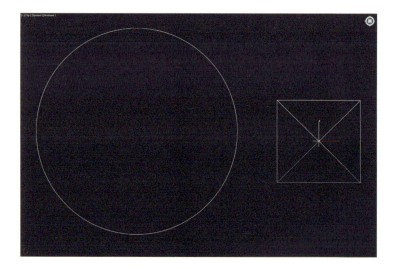

Figure 5.21
Creation of a *VRayLight* of random dimensions next to the object "light."

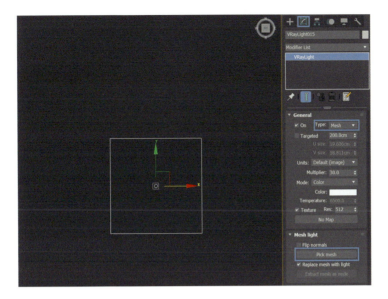

Figure 5.22

Selection of the *Mesh* type and activation of the *Pick mesh* command.

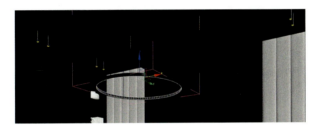

Figure 5.23

VRayLight takes the shape of the pendant.

Figure 5.24

Crop render of the pendant.

5.1 VRayLights

5.2 VRayIES

IES stands for the *Illuminating Engineering Society*, which has defined a file format for describing the distribution profile for a light. The information in those files, like the shape of the light cone or the steepness of the light falloff, are gathered through lab experiments and you can use the *IES* files in 3D applications to create accurate representations of light bulbs (Figure 5.25).

Go to the *Front* viewport, select the left track lights (layer *spot_track*), right-click, and choose *Isolate Selection*. Go to the *Command Panel*, click on the *Create* tab, and select *Lights*. Click on the *Photometric* drop-down list and select *VRay*. Choose *VRayIES* from the *Object Type* rollout (Figure 5.26).

Click and drag to place the *VRayIES* light below the spotlight. When you release the cursor, the light target is placed (Figure 5.27—left). Select both the light and the target, go to the *Top* viewport, and drag them to the correct position (Figure 5.27—right). Make instances and place them on the other spotlights. End the isolation and produce a crop render (Figure 5.28).

As mentioned in the *VRayLights*, make sure that the *VRayIES* lights are not placed in the object or in the *VRayLights* in order to cast light.

Figure 5.25
Different distribution profiles of lights.

Figure 5.26
Selection of *VRayIES*.

Figure 5.27

Creation of a *VRayIES* light in the *Front* viewport, its reposition in the *Top* viewport, and creation of instances for the remaining spotlights.

If you create a *VRayIES* light without loading an IES file, the light will have the shape of a sphere, as seen in the shadows it casts on the ceiling in Figure 5.28. To load an IES file, first you need to download it from the web. A site I could suggest is *LeoMoon.com*. Go to the menu *Asset Store* and from the *Shaders* download the *IES Lights Pack*. Go back to the *3ds Max* file, select one of the *VRayIES* lights and go to the *Modify* tab. Go to the *VRayIES Parameters* and click on the *None* button next to the *Ies file* (Figure 5.29).

Figure 5.28
Crop render to test the *VRayIES* lights.

From the *Open* dialog box, go to the *ies* folder you downloaded and select the *area-light.ies*. When you load the ies file, the preview of the light automatically changes. Produce a render and you will observe that the distribution of the light changes (Figure 5.30).

The lights are too bright, with the wall paneling getting washed out by their intensity. Select one of the *VRayIES* lights, go to the *Modify* tab, scroll down, and reduce the *Intensity value* to 10000. Moreover, instead of *Color*, select *Temperature* and set it to 4000 (Figure 5.31).

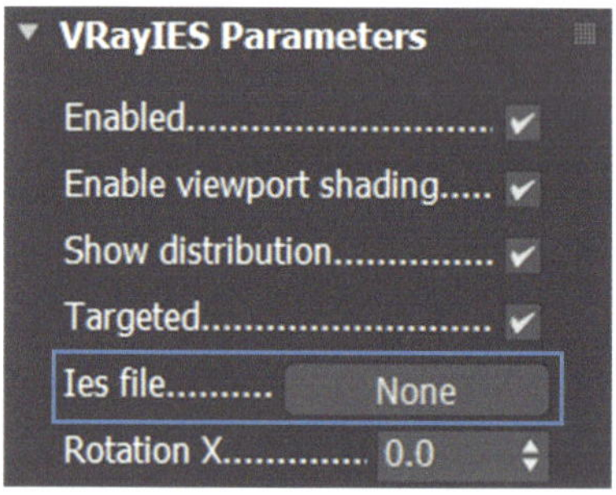

Figure 5.29
Use of the *None* button to load an *Ies* file.

Figure 5.30

Crop render showing the new light distribution.

Figure 5.31

Updated render with warm lights of 4000 K and reduced intensity.

5.2 VRayIES

Figure 5.32
Updated render with the new ies selection.

Try another ies file to see how the shape of the distribution changes depending on the ies selection. Thus, click on the *area-light* you loaded earlier in the *Ies file* field. The *Open* dialog box appears, select the *three-lobe-vee.ies,* and produce a render (Figure 5.32).

Note that in Figure 5.32, the shape of the light is barely visible on the wardrobe. A trick you can do in cases like this is the following. In general, the distribution of light can be visible if the *VRayIES* is placed close to a surface. Therefore, select the *VRayIES* lights, move them closer to the wardrobe, and lower them a few centimeters (Figure 5.33).

If you now render, the distribution of the light is visible on the wardrobe giving a nice pattern and shadows (Figure 5.34).

Clone the *VRayIES* of the left ceiling track and make *Instances* to the right ceiling track. Produce a render of the full viewport (Figure 5.35). If needed, adjust the *Exposure* layer. If you are satisfied with the result, you can produce the final render. Thus, go to the *Render Setup* dialog box (*F10*) and to the *Progressive Image sampler* rollout. Increase the *Render time* (or set it to 0.0) and reduce the *Noise threshold* to 0.005.

Figure 5.33
Reposition of the *VRayIES* lights closer to the wardrobe and a few centimeters lower.

Figure 5.34
Updated render after moving the *VRayIES* closer to the wardrobe.

Figure 5.35

Final render.

The final render in resolution 2000 × 1700 took approximately 1 hour. As mentioned in Chapter 2, you do not have to wait until the render resolves completely. Once you are satisfied with the result, you can press the *Stop* button and save the render. The render looks warm due to the warm artificial lighting. An easy way to adjust it is by adding the *White Balance* layer. For our example, set the *Temperature* to 5800 K. Moreover, add the *Curves* layer to increase the contrast of the render (Figure 5.36) (see Chapter 2).

Figure 5.36

Render after adding the *White Balance* layer and the *Curves* layer.

5.3 VRayLightMix

We explored the *VRayLightMix* in the previous chapter. However, this render element is extremely helpful when you are working on the artificial lighting of the scene.

VRayLightMix is introduced in *V-Ray 5*. If you are using an older version, I advise you to upgrade, since this feature is a great time saver.

While the render is being calculated or once it is cleared, go to the *Layers* in the *V-Ray Frame Buffer* window and select *Source:LightMix*. In the *Properties*, select *LightMix* and a list of all the lights in the scene appears (Figure 5.37). If the lights are *Instances*, only one of them will be included in the list.

It is useful, when you create a light, to rename it so that you can easily identify it in this list. If you cannot understand where each light refers to, simply click on the checkbox to turn the light off/on. If, for instance, you click on *VRayLight002*, the wall sconces turn off. Although you disable only *VRayLight002*, both wall sconce lights turn off, since they are *Instances*. If you turn off the *VRaySun001*, you no longer have light or shadows from the sun and the scene turns into a nighttime scene (Figure 5.38).

Figure 5.37

LightMix list showing all the lights in the scene.

Figure 5.38

Creation of a nighttime atmosphere by disabling the *VRaySun001*.

VRayLightMix allows you to control all the lights of a scene from one panel and quickly create different lighting scenarios. Every time you are satisfied with a result, click on the *Save current channel* button, to save the respective render. Apart from turning on/off lights via *VRayLightMix*, you can also control their intensity. For example, to reduce the intensity of the *VRayIES* lights, go to the *VRayIES012* and type 0.3 in place of 1 (Figure 5.39).

Moreover, you can control the color of the lights. If, for instance, you want to have a blue color on the wall lights, click on the color swatch next to the *VRayLight002* and choose a blue color (Figure 5.40).

When making adjustments in *LightMix*, always allow some time for *V-Ray* to recalculate and add the *Lens Effects*.

Finally, if you click on *LightMix* and you do not see a list of the scene lights, this means that you have not added *VRayLightMix* as a *Render Element* (Figure 5.41—left). In Chapter 2, we mentioned that you need to add mainly two elements in the *Render Elements* tab, the *VRayDenoiser* and the *VRayLightMix* (Figure 5.41—right). If you do not see it in the *Render Elements* list, click on the *Add* button and choose *VRayLightMix*.

Figure 5.39

Reduction of the intensity of the *VRayIES* lights.

Figure 5.40

Setting a new color for the light of the wall sconces.

5.3 VRayLightMix

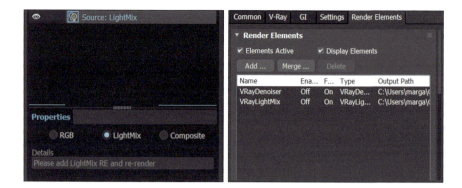

Figure 5.41

Selection of *VRayLightMix* from *Render Setup* > *Render Elements* > *Add* when there is a warning that *VRayLightMix* is not added.

6

Material Editor

The *Material Editor* provides functions to create and edit materials and maps. Therefore, it is important to get familiar with the *Material Editor* and how to use it before you start applying materials to the objects of a scene. There are two interfaces to the *Material Editor*, the *Compact* and the *Slate*. In this book, the *Slate Material Editor* is primarily used in various examples. However, the *Compact* is also described, so that you can choose which interface you prefer.

6.1 Slate Material Editor

Open a new *3ds Max* file and create a box and a sphere. Every object you create in *3ds Max* is automatically assigned a color. This color is not a material, but the color used for the object 2D and 3D representation in the viewports. To change this color, select the object, i.e., the box, go to the *Command Panel,* and click on the color swatch next to its name. The *Object Color* dialog box appears, where you can choose a new color (Figure 6.1).

To create a material, open the *Material Editor* by pressing the *M* key or clicking on the *Material Editor* button, ![icon], from the *Main Toolbar* (Figure 6.2).

As pointed out above, there are two interfaces to the *Material Editor*, the *Slate* and the *Compact Material Editor* (Figure 6.3). To switch between the two, click on the menu *Modes* in the *Material Editor* and select the interface you prefer (Figure 6.4).

The *Slate Material Editor* consists mainly of three windows, the *Material/Map Browser*, where you can browse for materials and maps, the *active View, View1*, where you can combine materials and maps, and the *Parameter Editor*, where you can control the materials and maps settings. To create a material using this editor, select the material from the *Material/Map Browser* and drag it to the *View1*.

DOI: 10.1201/9781003144786-6   127

Figure 6.1
Setting the *Object Color*.

Figure 6.2
Part of *Main Toolbar—Material Editor* button.

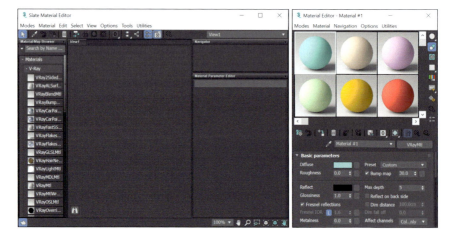

Figure 6.3
Slate Material Editor (left) and *Compact Material Editor* (right).

Figure 6.4

Modes menu.

For example, go to the *Material/Map Browser*, go to *Materials*, then *V-Ray*, and drag and drop the *VRayMtl* in *View1* (Figure 6.5).

If you do not see the *VRayMtl* in the *Material/Map Browser*, open the *Materials* category and then the *V-Ray* sub-category. Otherwise, simply type

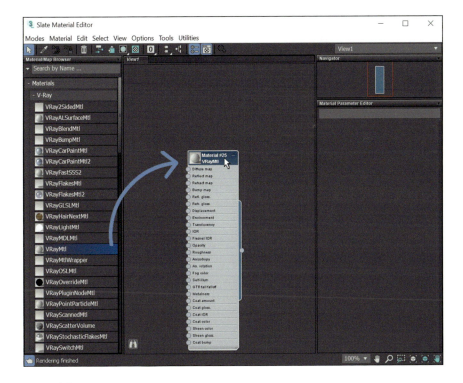

Figure 6.5

Steps to create a *VRayMtl* using the drag and drop method.

6.1 Slate Material Editor

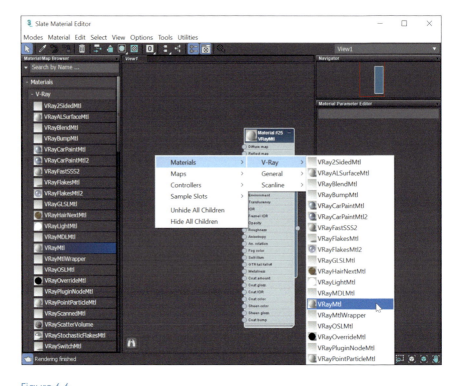

Figure 6.6

Steps to create a *VRayMtl* using the right-click.

VRayMtl in the search bar of the *Material/Map Browser*. Another way to create a *VRayMtl* is to right-click in the *active View* and select *Materials > V-Ray > VRayMtl* (Figure 6.6).

Click and drag to reposition a material in the *active View*. Moreover, you can zoom in, zoom out, or pan in the *active View*. To delete a material, select it and hit the *Delete* button from the keyboard.

In *View1*, the material appears as a node. A node consists of the following parts (Figure 6.7):

1. The title bar that shows a small preview of the material/map, followed by the name of the material/map, and then the material/map type.
2. The slots that show components of the material/map.
3. To the right of the entire node, there is a small circle called output socket. When you click on it and drag, a red wire appears. Use this to wire/assign the material to an object (or a map to a material).
4. To the left side of each slot, there are small circles called input sockets. Use them to wire a map.

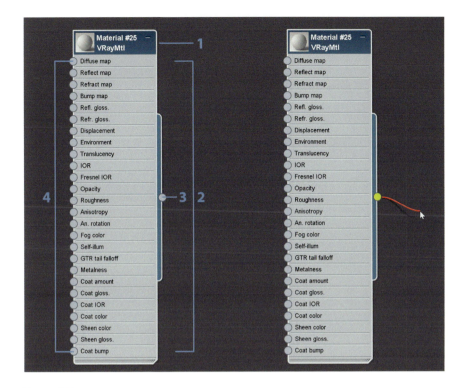

Figure 6.7

VRayMtl material node (left). Use of the output socket to assign the material to an object (right).

You can collapse a node to hide its slots or expand it to show the slots by clicking on the −/+ icon at the top right corner of the title bar. You can also resize a node horizontally, by clicking the symbol at the bottom right corner of the node. Enlarge the material preview by double-clicking the preview in the node title bar. To reduce the preview, double-click the preview again (Figure 6.8). Double-click anywhere on the node and the material settings open in the *Parameter Editor*. When the node parameters are displayed in the *Parameter Editor*, in the *active View* the material node appears with a dashed border.

To assign the *VRayMtl* to the box, click on the output socket, drag, and release the cursor on the box (Figure 6.9). The box acquires a medium gray color. Do not get confused during this process and try to drag the material node on the object. You need to click on the output socket, hold down the left mouse button, and release it on the object.

Another way to assign a material to an object is the following. Select the sphere, click on the second *VRayMtl* node you created to activate it, and click on

6.1 Slate Material Editor

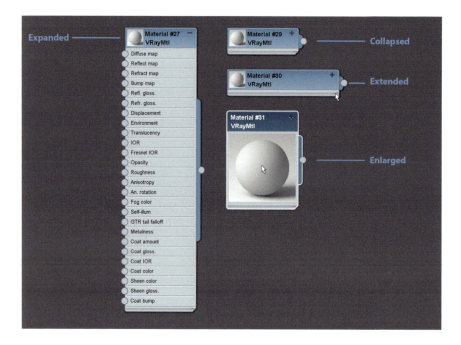

Figure 6.8
VRayMtl node expanded, collapsed, extended, and enlarged.

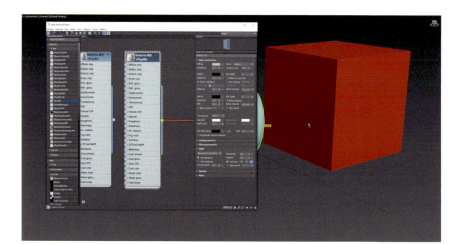

Figure 6.9
The drag and drop method to assign a material.

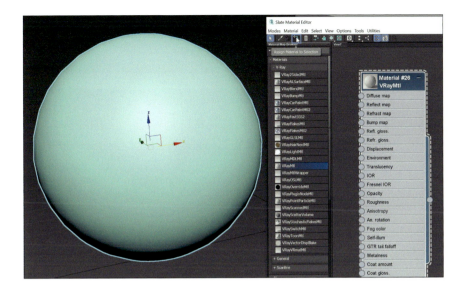

Figure 6.10

The *Assign Material to Selection* method.

the fourth button in the *Material Editor*, the *Assign Material to Selection*, ![icon]. The material assigns to the selected object (Figure 6.10).

The *VRayMtl* is the typical material that *V-Ray* offers and by default is a medium gray color. Go to the *Parameter Editor*. At the top, you see the material name. *V-Ray* automatically names every new material you create. Rename it to *box*. The material node in *View1* is automatically renamed (Figure 6.11).

Go to the *Basic Parameters* rollout and from the *Diffuse* section select the color of the material. Click on the *Diffuse* color and the *Color Selector: diffuse* opens. Click in the *Hue* panel to select a color, say a red hue, and use the *Whiteness* bar to adjust how dark or bright the color will look (Figure 6.12). Any changes you make can be seen in real-time in the camera viewport after assigning the material to the object.

If you know the RGB value, you can type it in the *Red, Green,* and *Blue* fields.

Figure 6.11

Renaming an object.

6.1 Slate Material Editor 133

Figure 6.12
Ways to set a *Diffuse* color.

Figure 6.13
Steps to add a texture.

To add a texture, click on the small box next to the color swatch. The *Material/Map Browser* opens. Go to *Maps,* click on *V-Ray* to open (if needed), choose *VRayBitmap,* and click *OK* (Figure 6.13).

A map node appears in the *active View*, the *VRayBitmap,* and is wired to the *Diffuse map* input socket. If the *VRayBitmap* overlaps other *VRayMtl* nodes, as in Figure 6.14, click and drag the nodes to reposition them. The wire helps you to understand that the material has a bitmap assigned to the *Diffuse map*. Moreover, in the settings, next to the *Diffuse* color, you see the uppercase letter *M* (map).

Another way to add a *VRayBitmap* is to right-click in the *View1* and choose *Maps > V-Ray > VRayBitmap*. Then, connect the *VRayBitmap* to the *Diffuse map* input socket (Figure 6.15).

To delete a node, click on it and press the *Delete* button from the keyboard. If we double-click on a node, we can see the respective parameters in the *Parameter Editor*. Thus, if you double-click on the *VRayBitmap* node, its parameters open in the *Parameter Editor*. Go to the *Parameters* rollout and click on the *3 dots* button

Figure 6.14
Overlapping nodes (left) and repositioned nodes (right).

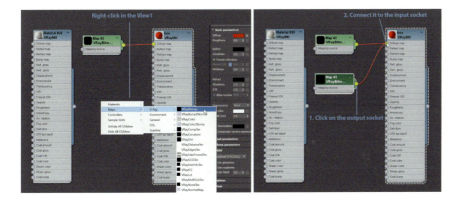

Figure 6.15
Steps to add a texture using the right-click.

next to the *Bitmap* field (Figure 6.16). The *Choose HDR image* dialog box opens. Go to the folder *Chapter 6* and select a texture, for example, the *fabric panel*.

The preview of the *VRayMtl* in the *active View* changes to the selected texture, but in the camera viewport, the box still looks red. For a texture to be visible in the camera viewport, click on the *VRayBitmap* node and then click on the *Show Shaded Material in Viewport* button, ▣, in the *Material Editor* (Figure 6.17).

Even if you forget to enable the *Show Shaded Material in Viewport* command, *V-Ray* will render the bitmap.

6.1 Slate Material Editor 135

Figure 6.16
3 dots button used to load a bitmap.

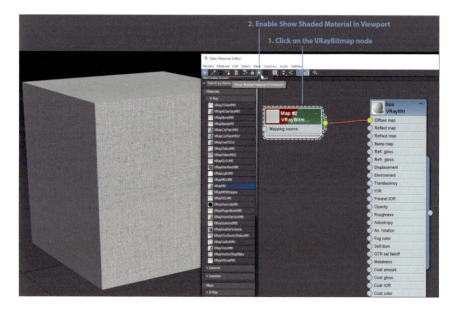

Figure 6.17
Show Shaded Material in Viewport command.

To control the scale of the texture, go to the *VRayBitmap* parameters, to the *Coordinates* rollout, and adjust the *Tiling*. If you make the tiling 2 by 2, the texture becomes twice as dense, while if you make the tiling 0.5 by 0.5, the texture becomes twice as big (Figure 6.18).

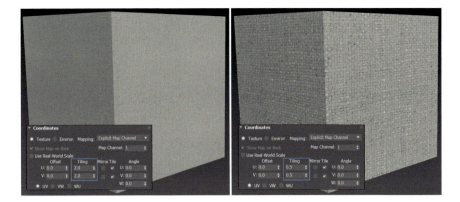

Figure 6.18

Tiling 2 × 2 (left) and 0.5 × 0.5 (right).

To change the texture, double click on the *VRayBitmap* node and go to the *Parameters* rollout. Click on the *3 dots* button and choose another texture, for example, the *wood_b*. To change the orientation of a texture, go to the *Coordinates* rollout in the *VRayBitmap* parameters and to the *Angle* section. If you type 90 in the *W* field, the texture rotates by 90° (Figure 6.19).

Another way to change the scale and the orientation of a texture is by using the *UVW Map* modifier, which is explained in Chapter 7.

To sum up, for the creation of materials/maps, there are two basic procedures described above. The first one is to use the *Material/Map Browser* of the

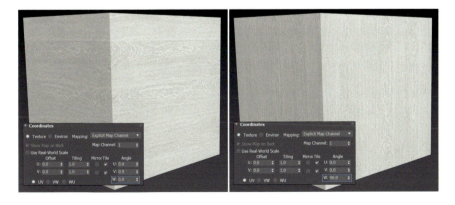

Figure 6.19

Control the direction of the texture in the *Angle* section.

6.1 Slate Material Editor

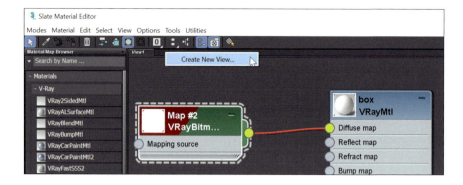

Figure 6.20

Create New View command.

Material Editor. Go to the *Material/Map Browser* and choose *Materials > V-Ray*. Click on *VRayMtl* and drag it in *View1*. Then go to the *Material/Map Browser* and choose *Maps > V-Ray*. Click on *VRayBitmap* and drag it in *View1*. Finally, connect the *VRayBitmap* to the *Diffuse map* input socket. The second way is to right-click in the *active View, View1*. More specifically, right-click in *View1* and choose *Materials > V-Ray > VRayMtl*. Then right-click one more time and choose *Maps > V-Ray > VRayBitmap*. Finally, connect the *VRayBitmap* to the *Diffuse map* input socket.

Regarding the *active View*, you can create several *Views*. To do so, right-click next to the *active View, View1,* and choose *Create New View* (Figure 6.20). In this way, you can keep the materials of the scene organized in different views.

6.2 Compact Material Editor

Open the *Material Editor (M)*, click on the menu *Modes* and choose *Compact*. The *Compact Material Editor* is a smaller, and as its name denotes, a more compact interface than the *Slate Material Editor*. The interface consists of a *menu bar* (1) at the top, the *sample slots* (2), and the *toolbars* (3) located at the bottom and the side of the sample slots. Moreover, it has several *rollouts* (4) that include the settings of the materials (Figure 6.21).

By default, each slot shows a *VRayMtl* in different colors. To assign a material to an object, simply drag and drop the slot on the object (Figure 6.22).

Another way to assign a material to an object is to select the slot, select the object, and click on the *Assign Material to Selection* command, from the *Material Editor* toolbar (Figure 6.23).

When a material is assigned to the scene, there are white triangles at the corners of the preview window. White stroke triangles mean that the material is assigned to the scene. Solid white triangles mean that the material is assigned to the selected object (Figure 6.24).

Figure 6.21

Compact Material Editor.

To create a material, select an empty slot, for instance, the second slot from the upper row, and make sure that the *VRayMtl* type is selected. Otherwise, click on the material type field to open the *Material/Map* Browser, select *VRayMtl* and click *OK* (Figure 6.25).

The *rollouts* include the same settings as the ones in the *Slate Material Editor*. Thus, if you want to load a texture, click on the small box next to the *Diffuse* color in the *Basic Parameters* rollout. The *Material/Map Browser* appears, select *VRayBitmap*, and press *OK*.

You are automatically transferred to the *VRayBitmap* parameters. Click on the *3 dots* button and choose a texture, for example the *fabric panel*. As in the

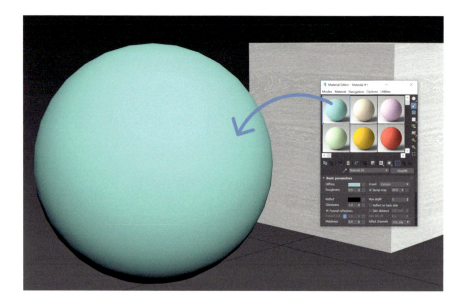

Figure 6.22

The drag and drop method to assign a material.

Slate Material Editor, to see the texture in a viewport, click on the *Show Shaded Material in Viewport* button, . To go back to the *VRayMtl* settings, click on the *Go to Parent* button, . To go to the *VRayBitmap* parameters, click on the button with the *M* letter next to the *Diffuse* color (Figure 6.26).

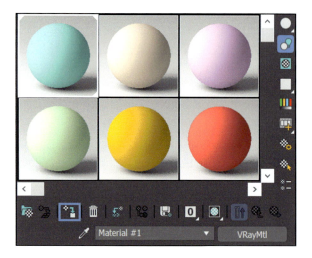

Figure 6.23

The *Assign Material to Selection* method to assign a material.

Figure 6.24

The two types of triangles: white stroke and solid white.

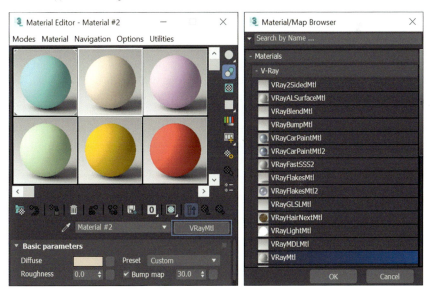

Figure 6.25

Selecting the *VRayMtl* type.

Figure 6.26

VRayBitmap settings button (left) and *Go to Parent* button (right).

6.2 Compact Material Editor 141

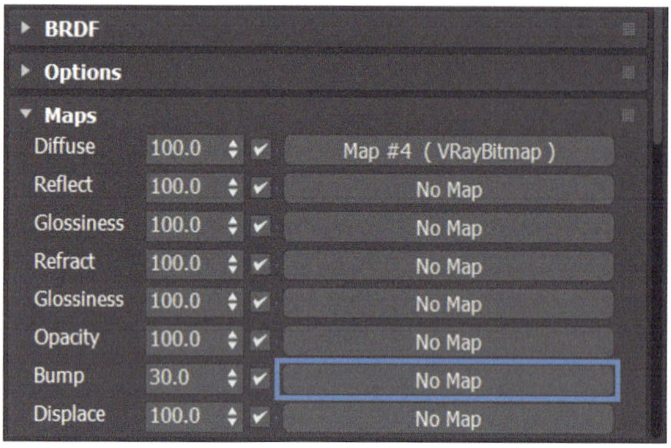

Figure 6.27

No Map button.

To add a *Bump map*, go to the *VRayMtl* parameters window, scroll down in the parameters to find, and open the *Maps* rollout. Click on the *No Map* button next to the *Bump* setting (Figure 6.27). From the *Material/Map Browser* choose *VRayBitmap*, click on the *3 dots* and select a texture, for example, the *fabric_bump*.

As before, use the *Go to Parent* button to return to the *VRayMtl* settings. The *Maps* rollout is very helpful to see in which attributes you have assigned maps. Alternatively, check for the *M* letter next to every setting that you have applied the map. When the *M* letter is uppercase, the map is assigned and active. If it is a lowercase *m*, the map is assigned, but inactive. To make a map inactive, go to the *Maps* rollout and disable the checkbox next to the respective setting (Figure 6.28). When a map is inactive, the object will be represented by the *Diffuse* color.

If you want to *Copy* a material, drag and drop it in a new slot.

By default, there are six sample slots (3 × 2) visible and you can scroll down to see more. In total, there are 24 slots. You can click on the *Options* menu and select

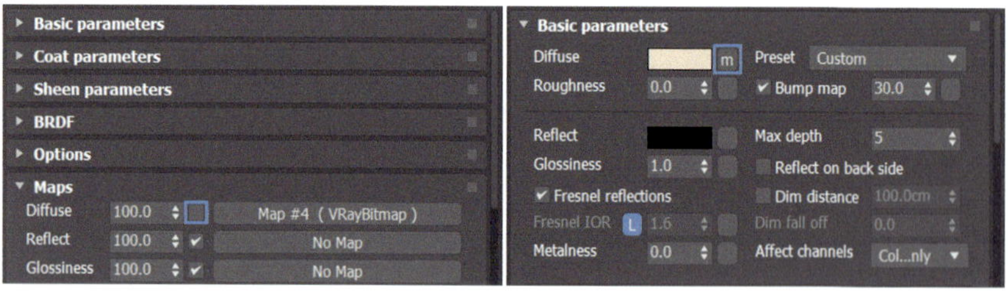

Figure 6.28

Inactive map.

142   6. Material Editor

Figure 6.29

Steps to reset a slot.

another cycle, for example 5 × 3 slots. If you click one more time on the cycles, you see all 24 slots and the scrolling is disabled. Click one more time to go back to the 3 × 2 previews.

Every slot can host only one material and in total there are 24 slots. This means that if in your project you have more than 24 materials, you need to start overriding the slots, i.e., reset them to their default settings and then create the new material. Note that resetting the *Compact Material Editor* sample slots does not remove the materials already used in the scene, it simply replaces the selected slot with a default *VRayMtl*. As an example, select the second slot and click on the trashcan button, the *Reset Map/Material to Default Settings*, . The *Reset Mtl/Map Params* window appears, where you can choose if you want to delete the material both from the scene and from the *Material Editor* or if you only want to delete the material from the *Material Editor*. Choose the second option and press *OK*. The material is overridden and the default middle gray *VRayMtl* takes its place (Figure 6.29). But the old material in the scene remains intact.

To modify a material that is applied in the scene and is not in the *Material Editor*, you can select a sample slot and click on the *Pick Material from Object* command, (Figure 6.30). Then click on the object. Its material is transferred to the selected slot. We will see this command more in depth in Chapter 8.

Figure 6.30

Pick Material from Object command.

6.2 Compact Material Editor 143

6.3 Slate vs Compact Material Editor

The *Slate Material Editor* uses nodes and wiring to graphically display the structure of materials. The *Compact Material Editor* is a smaller, more compact interface that uses previews of the materials. Despite the different appearance, all the commands are the same, either you use the *Compact* or the *Slate Material Editor*.

My advice is to use the *Slate Material Editor*, since its expanded interface allows you to have a better overview of your materials. More specifically, in the *active View*, the wires help you to see the textures you have loaded in each setting. You can see a preview of all the materials and maps and you can easily identify which are instances and which are copies. So, modifying a material is much quicker in the *Slate Material Editor*. Another advantage of the *Slate Material Editor* is that in the *active View*, you can create as many materials as needed and you do not have to count the empty slots as in the *Compact Editor*. Finally, the fact that you can create multiple views in the *Slate Material Editor* gives you the option to organize the materials of the scene.

The main disadvantage of the *Slate Editor* is its size and the fact that you constantly need to move it around, since it covers big part of the viewports. But the size is adjustable. While you are moving the *Slate Material Editor*, you might accidentally relocate one of its windows (Figure 6.31). To dock a floating window, double click on the window title bar and it will dock in its default location. Otherwise drag it in the *Slate Material Editor* and arrows will appear that show

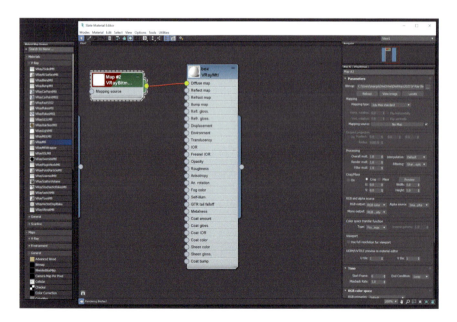

Figure 6.31

Material/Map Browser floating outside the *Slate Material Editor*.

Figure 6.32

Docking a window in the *Slate Material Editor*.

the possible positions to dock (Figure 6.32). Drop it on one of the arrows and it will attach to that side.

Now that you know how to create, apply, and edit a *VRayMtl*, you are ready to start applying materials to the objects of the scene, an issue examined in the next chapter.

7
Materials

Materials are one of the most important parts of the process of achieving a photorealistic result. Usually, for one single material we need to apply a set of properties to give the impression that an object is soft and shiny, like leather, or clean and reflective, like glass. In this chapter, we examine how to set all the necessary settings in a material to give realism to a scene. Moreover, we explore the use of *Asset Tracking* to link any missing textures from the scene and, finally, the use of the *UVW Map* modifier to adjust the way a texture is projecting on an object.

7.1 The Scene Materials

Open the file *Chapter 7.max*. In this *3ds Max* scene, there is a *VRayPhysicalCamera* with *EV* set to 9, a *VRaySun*, and *VRayLights*. A white color (RGB 220) is assigned to all the objects in the scene, since in this chapter, we will explore the materials. As in the previous chapters, set to do 5-minute test renders. If you press the *Render* button, the render will be burnt by the lights in some areas. Add the *Exposure* layer to adjust the *Highlight Burn* value as well as the *White Balance* and the *Curves* to adjust the render (Figure 7.1).

Although there is a *VRaySun* in the scene, it does not cast light in the interior. That is because the sheer has a solid color, i.e., it does not have opacity, and so it blocks the sunlight from entering. Since the sheer material will be examined later in this chapter, what you can do now to let the sun in is to select the *VRaySun001* and go to the *Modify* tab. Go to the *Options* section and click on the *Exclude* button. The *Exclude/Include* dialog box appears that shows on the left column all the objects of the scene. Select the *sheer* and click on the right arrow to move the object to the right column (Figure 7.2). In this way, the *VRaySun* excludes the sheer. Click *OK* and produce a render (Figure 7.3).

DOI: 10.1201/9781003144786-7

Figure 7.1
Render with *Highlight Burn*: 0.40, *Contrast*: 0.10, *White Balance*: 5000, and *Curves* adjusted.

Figure 7.2
Steps to exclude the sheer from the *VRaySun*.

Figure 7.3

Render with the sheer excluded. *Exposure*: −3.00 and *Highlight Burn*: 0.20.

It is seen that the sun enters the room casting light and shadows and so the interior gets much brighter. You need to re-adjust the *Exposure* layer and the *Exposure* value since the render is now too bright (Figure 7.3).

Once you apply the correct material to the sheer, you need to go to the *VRaySun* parameters and include back the sheer.

When you work with the materials, you need to keep adjusting the *Exposure* layer to control the brightness of the render.

Now we can proceed to create some basic materials:

7.1.1 Fabric

The first material we will create is the fabric for the panels on the left and right sides of the bed. We will basically repeat the steps we did in Chapter 6. Thus, press M, create a *VRayMtl* and rename it to *fabric panel*. Create a *VRayBitmap* and wire it to the *Diffuse map*. Go to the *VRayBitmap* node parameters, click on the *3 dots* button, go to the folder *Chapter 7*, and choose the image *fabric panel*. Click on the *VRayBitmap* node and click on the *Show Shaded Material in Viewport* command in the *Material Editor*. Click on the output wire, drag, and drop it on the panels, layer *wall panels* (Figure 7.4). Produce a render (Figure 7.5).

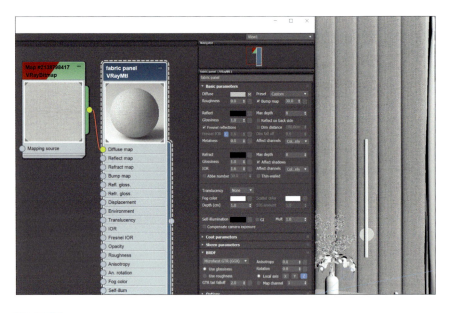

Figure 7.4

Fabric *VRayMtl* applied to the *wall panels*.

Figure 7.5

Render with the fabric material applied.

Figure 7.6

Crop method with *Output Size* set to 3000 × 2250.

When you work with materials, it is good to do crop renders to focus on the material you are creating. Moreover, it is good to render in higher resolution, so that you can clearly see how the material renders.

Go to the *Render Setup* dialog box and to the *Output Size*. Make sure the lock is enabled next to the *Image Aspect*, which is equal to 1.3333 and make the *Width* 3000 pixels. Go to the *Area to Render*, choose *Crop*, and select a small area in the camera viewport to render (Figure 7.6). Now you can clearly see how the fabric material looks on the wall panels (Figure 7.7).

The 3000 pixels width refers to the width of the full viewport. Since you do a crop render, the width of the render will be less than 3000 pixels. You can see the dimensions of the render in the title bar of the *V-Ray Frame Buffer* window.

7.1.2 Wood

The next material we will create is the wood that will be applied to the slats, the wall behind them, the nightstands, and the wardrobe. Open the *Material Editor*. Using the *Pan* command create some space in *View1*, right-click and choose *Materials > V-Ray > VRayMtl*. Double-click on the node and rename it to *wood*. Create a *VRayBitmap* and wire it to the *Diffuse map*. Go to the *VRayBitmap* node parameters, click on the *3 dots* button, go to the folder *Chapter 7*, and choose the image *wood_d*. Click on the *VRayBitmap* node and then on the *Show Shaded Material in Viewport* command in the *Material Editor*. Press *H* to open the *Select from Scene* dialog box and select the following layers: *ceiling slats, nightstands, slats, wardrobe,* and *wall wood* and press *OK* (Figure 7.8—left). Click on the wood

Figure 7.7
Crop render of the fabric.

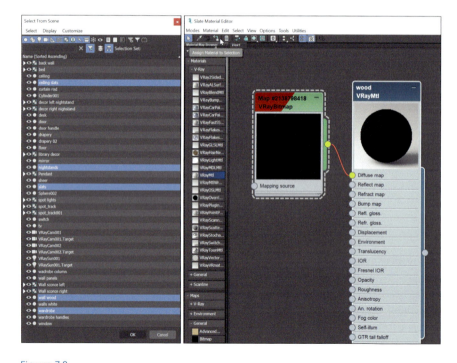

Figure 7.8
Layer selection and activation of the *Assign Material to Selection* command.

VRayMtl in the *Material Editor* to activate it and click on the *Assign Material to Selection* command (Figure 7.8—right).

To see the layer *wardrobe wood panels*, you need to click on the arrow next to the *wardrobe* to see the group elements.

The *Slate Material Editor* is a quite big dialog box, so you need to keep moving it around while you are working on the materials, so that it does not block the viewports. Go to the *Render Setup* window, adjust the *Region* box to include the nightstand, and produce a render (Figure 7.9).

To change the direction of the wood texture, double-click on the *VRayBitmap* node to see its parameters and go to the *Coordinates* rollout. Go to the *Angle* section and type 90 in the *W* field (Figure 7.10).

The nightstand does look like wood, but it is flat. For a material to render realistic, we need to apply more attributes than just assigning a texture to the *Diffuse map*. One of these attributes is the reflections. Double-click on the *wood VRayMtl* to open its parameters and go to the *Basic Parameters* rollout. Click on the *Reflect color* and go to the *Whiteness* column. When the *Reflect* color is black, the material is matte because it does not reflect. The lower you slide toward white, the more reflective the material becomes. If you set it to white, you get a very reflective surface (Figure 7.11). Make the *Reflect* color white and render (Figure 7.12).

Now set the *Reflect* color to RGB 44, which is a medium gray color, and produce a render. The wood is still reflective, but not as much as it was before (Figure 7.13).

If you take a close look at the reflections, you will notice that there are mirror-like reflections, i.e., you can clearly understand the objects reflected on the nightstand. A material can have different types of glossiness. It can have a polished finish, where you have clean reflections, or it can have a satin finish, where the

Figure 7.9
Crop render of the right nightstand.

7.1 The Scene Materials

Figure 7.10
Wood texture rotated by 90°.

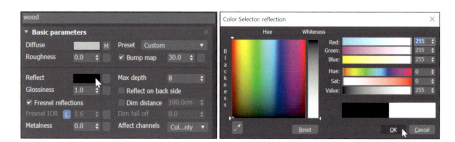

Figure 7.11
Steps to set the *Reflect* color to white.

Figure 7.12
Reflect color of wood set to white.

154 7. Materials

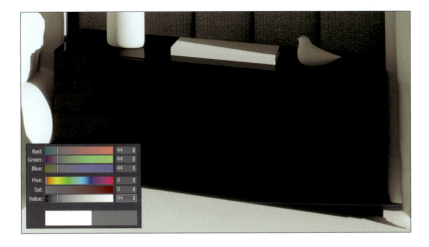

Figure 7.13

Crop render with the *Reflect* color set to RGB: 44.

reflections are blurred. In the latter case, the material is still reflective, but you cannot understand what is reflected on it. How blurred the reflections will be is controlled by the *Reflect Glossiness* value. Go to the *wood VrayMtl* parameters. When the *Glossiness* in the *Reflect* section is 1, you have mirror-like reflections. The lower you go, the blurrier the reflections you get. Try 0.8 and render. In Figure 7.14, the wood is glossy, but you cannot understand anymore what is reflected on it.

Instead of a color, you can assign a texture to the *Reflect* setting. The philosophy is the same as applying a color. You need to choose a black and white or

Figure 7.14

Reflect RGB: 44 and *Reflect Glossiness*: 0.8.

7.1 The Scene Materials

Figure 7.15

Reflect map wired with a grayscale texture and rotation of the texture by 90°.

a grayscale texture. Where *V-Ray* detects black it makes that part of the object matte, where it detects white it makes it very reflective, and where it detects gray values, it adjusts them accordingly. So, when you use a texture in the *Reflect map*, you achieve a variety of reflections.

As an example, open the *Material Editor*, create a *VRayBitmap* and connect it to the *Reflect map*. Click on the *3 dots* button next to the *Bitmap*, go the folder *Chapter 7*, and choose the image *wood_r*. Select the *Reflect map VRayBitmap* node and click on the *Show Shaded Material in Viewport* to see how the texture applies to the objects (Figure 7.15). As before, you need to go to the *Angle* section and type 90 in the *W* field to rotate the texture by 90°. Produce a render (Figure 7.16).

Figure 7.16

Render with *Reflect* set using a grayscale texture and *Reflect Glossiness*: 0.8.

If you compare Figure 7.10 to Figure 7.16, you can see that when you add some reflections to the wood material and adjust the *Reflect Glossiness*, you get a nice wooden finish.

7.1.3 Glass

Create a new *VRayMtl*, double-click on the node to open its parameters, and rename it to *glass*. In the *Basic parameters* rollout, next to the *Diffuse* color is the *Preset* setting. Click on the *Custom* button and a drop-down menu appears that contains some pre-made materials that *V-Ray* offers. Select *Glass* (Figure 7.17).

If you forget to double-click on the node and go directly to the *Parameter Editor*, you will change the settings of the previously active material. You must always first double-click on the node you want to adjust.

Apply the material to the vase on the nightstand. The vase becomes transparent since it is clear glass and is barely visible in the viewport. Produce a render (Figure 7.18).

Open the *Material Editor* and double-click on the *glass VRayMtl* node to open its settings. So far, we have seen the *Diffuse* and *Reflect* settings. Another interesting parameter is the *Refract*. *Refract* has the same philosophy as the *Reflect* setting, which means that when the *Refract* color is black, the object is solid, i.e., there is no opacity. If the *Refract* color is white, the object is fully transparent, while using values between black and white, you control the transparency of the material. Thus, if instead of white you make the *Refract* color a light gray, RGB 143, you will be able to see the vase in the viewports and if you render, the vase is semitransparent (Figure 7.19).

Figure 7.17

Glass selection from the *Preset* list.

Figure 7.18
Preview (left) and render (right) of the vase using the *Glass* preset.

Figure 7.19
Preview (left) and render (right) of the vase with *Refract* RGB 143.

Figure 7.20

Diffuse: 3/5/1, *Reflect*: 233, and *Refract*: 181.

To make a colored glass, set the *Diffuse* to the desired color, the *Reflect* color to white since the glass is a reflective material and play around with the *Refract* color, depending on how transparent you want the glass to look. More specifically, to make the vase a green glass, set the *Diffuse* color to RGB 3/5/1, the *Reflect* color to white, and the *Refract* color to 181 (Figure 7.20).

7.1.4 Metal

The next material we will create is the black metal of the light fixtures and the wardrobe handles. Create a new *VRayMtl*, rename it to *metal*, and from the *Preset* drop-down menu choose *Plastic*. Double-click on the *VRayMtl* node to open its settings. Change the *Diffuse* color to a dark gray, RGB 2, and set the *Reflect Glossiness* to 0.75 to have blurred reflections (Figure 7.21).

Apply the *metal VRayMtl* to the following layers: *Pendant, spot_track, spot_track001, Wall sconce left, Wall sconce right, wardrobe handles*, and do a crop render at the pendant (Figure 7.22).

Produce a render of the full viewport to see what you get so far. First, go to *Render Setup* and change the *Output Size* to 1200 × 900. Then go to the *Area to Render* and choose *View* (Figure 7.23).

7.1 The Scene Materials   159

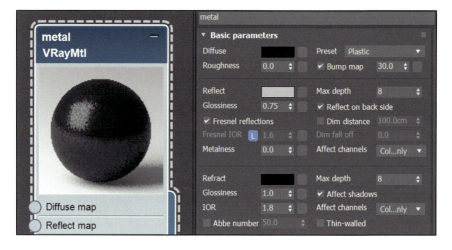

Figure 7.21

VRayMtl using the *Plastic* preset with *Diffuse* RGB: 2 and *Reflect Glossiness*: 0.75.

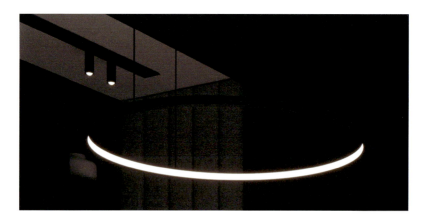

Figure 7.22

Crop render of the pendant.

It is seen that as you continue to apply materials and especially when these materials are of dark hues, the darker the render becomes. So, you should continue to adjust the *Exposure* layer, while working with the materials.

If you now notice the wood in Figure 7.23, you see that the material does not really look like wood. It has a more glossy brown finish and you can no longer see the wood knots, while they are visible when you are doing crop renders. This happens because of the combination of low rendering settings (5-minute rendering time) and the *VRayDenoiser*, which blurs the textures to cover the scene noise. Once you increase the *Render time,* the wood texture will render properly.

Figure 7.23

Bedroom render.

7.1.5 Drapery and Sheer

In this project, we have two types of curtains, the drapery at the sides and the sheer at the center. The drapery is basically a solid fabric and you can create it following the same steps you did to create the fabric of the wall panels. More specifically, create a new *VRayMtl* and rename it to *drapery*. Create a *VRayBitmap*, connect it to the *Diffuse map*, and load the image *fabric beige* from the folder Chapter 7. Assign the material to the objects *drapery* and *drapery 02*.

The sheer has a semitransparent fabric. Create a new *VRayMtl* and rename it to *sheer*. Apply a *Diffuse* color with an RGB 233, and keep the *Reflect* color black, since you do not want the fabric to have any reflections. Click on the *Refract* color and set it to a light gray color, RGB: 126 (Figure 7.24). Apply the material to the object *sheer*.

> If you find it hard to drag and drop the material to the drapery or the sheer, use the *H* key to select the layer *sheer* and then use the *Assign Material to Selection* command from the *Material Editor*.

Before you produce a render, select the *VRaySun*, go to its properties in the *Modify* tab and click on the *Exclude* button. Select the *sheer* from the right list and click on the left arrow to move the sheer back to the *Scene Objects* list (Figure 7.25).

7.1 The Scene Materials

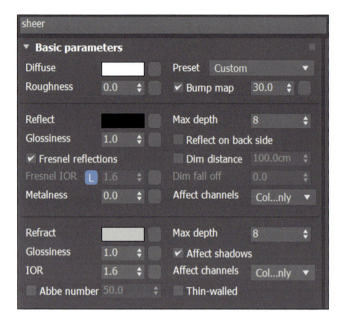

Figure 7.24
Sheer values set to *Diffuse*: 233, *Reflect*: 0, and *Refract*: 126.

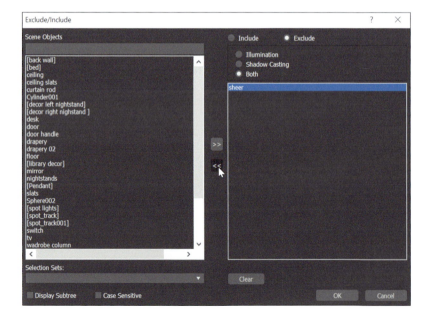

Figure 7.25
Steps to move the sheer back to the scene objects.

Figure 7.26

Render with the sheer included in the *VRaySun*, *Exposure*: −0.70, and *Temperature*: 5500.

Note that at the beginning of this chapter you excluded the sheer from the *VRaySun*, because it prevented sunlight from entering. However, now you applied the correct sheer material and, therefore, you should re-include the object. Produce a render (Figure 7.26).

It is seen that by applying a semitransparent material to the sheer, the sunlight is more subtle, does not have as strong shadows as it had before, and the waves of the sheer cast shadows on the floor and on the bed giving more realism and a nice effect to the render.

7.1.6 Leather

Create a new *VRayMtl* and rename it to *leather*. Create a *VRayBitmap* and connect it to the *Diffuse map*. Load the texture *leather* from the folder *Chapter 7*. Assign the material to the bed frame, click on the *VRayBitmap,* and then on the *Show Shaded Material in Viewport* command to see the texture in the camera viewport (Figure 7.27). Produce a crop render (Figure 7.28).

To give the feeling of leather, you also need to add a *Bump map. Bump* creates the illusion of depth on a surface. This technique uses grayscale images and the parts of the image that are bright toward white get pulled out of the surface, whereas the parts of the image that are dark toward black get pushed into the surface creating the illusion of depth. Thus, create a new *VRayBitmap*, load the

7.1 The Scene Materials 163

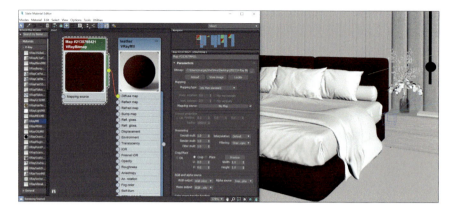

Figure 7.27

Leather *VRayMtl* applied to the bed frame.

Figure 7.28

Crop render of the bed frame.

image *leather_bump*, and connect it to the *Bump map*. Click on the *VRayBitmap* node and then on the *Show Shaded Material in Viewport* button to see how the texture applies to the bed (Figure 7.29).

The scale of the texture is too big, so double-click on the *VRayBitmap* node to open its parameters, go to the *Coordinates* rollout, make the *Tiling* 5 by 5, and produce a crop render (Figures 7.30 and 7.31).

To control the strength of the *Bump* effect, go to the *Maps* rollout in the *VRayMtl* parameters. Next to the *Bump* is a numeric field (Figure 7.32). By default, the *Bump* is set to 30%. If you increase this value, the bump effect will be stronger, while decreasing the value will reduce the bump effect (Figure 7.33). Make the *Bump* value for the leather of this scene 10.

Figure 7.29
Preview of the *Bump map*.

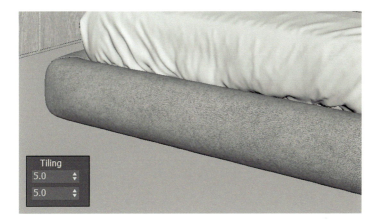

Figure 7.30
Tiling of *Bump map* set to *U*: 5 and *V*: 5.

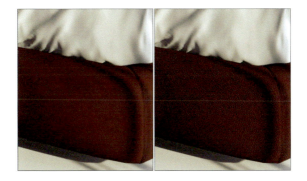

Figure 7.31
Leather before (left) and after (right) adding a *Bump map*.

7.1 The Scene Materials 165

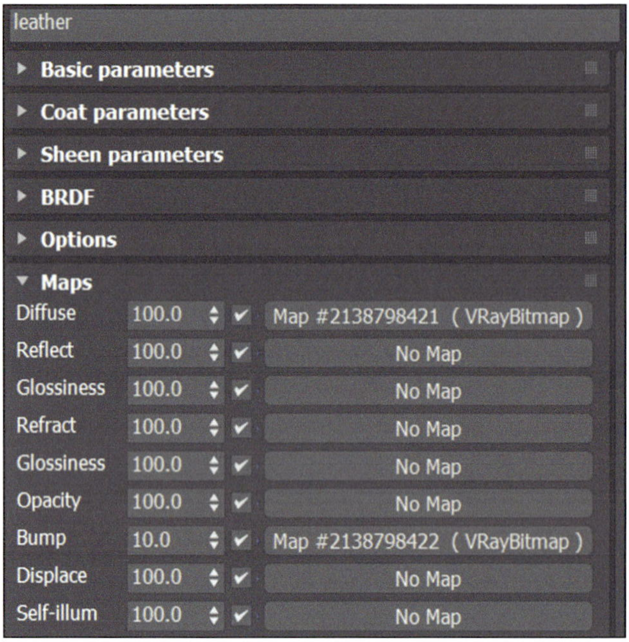

Figure 7.32
Maps rollout.

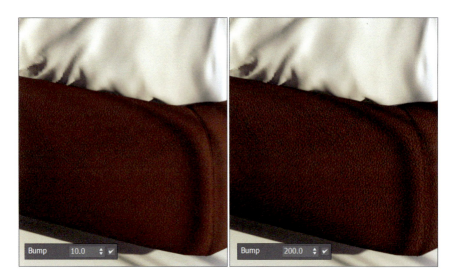

Figure 7.33
Bump set to 10 (left) and 200 (right).

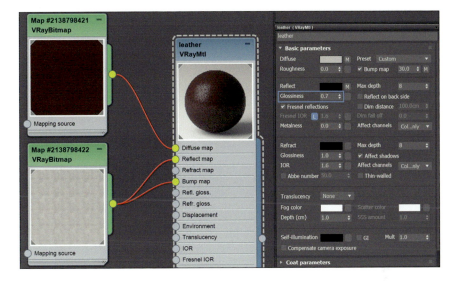

Figure 7.34

Reflect map and *Bump* map wired with the same grayscale texture and *Reflect Glossiness*: 0.7.

Finally, add some reflections to the leather material. Instead of creating a new *VRayBitmap*, you can wire to the *Reflect* map the same node that is wired to the *Bump* map. Moreover, make the *Reflect Glossiness* 0.7 to have blurry reflections (Figures 7.34 and 7.35).

Figure 7.35

Render of leather after adding a *Reflect* map.

7.1 The Scene Materials 167

7.1.7 Carpet

Create a new *VRayMtl*, rename it to *carpet* and apply it to the bedroom floor (layer *floor*). Create a *VRayBitmap* and load the texture *carpet* from the folder *Chapter 7*. Connect the *VRayBitmap* both to the *Diffuse* map and the *Bump* map, and produce a render (Figure 7.36).

Produce a render of the full viewport to see what you get so far (Figure 7.37). You can increase the *Render time* to get a cleaner image. Do not forget to adjust the *Exposure* layer, if needed.

Figure 7.36

Crop render of the carpet.

Figure 7.37

VRayCam001 viewport render.

7.1.8 Book (Multi/Sub-Object Material)

The next material we will create is the book material. If you select the book on the right nightstand, the full group is being selected (layer *decor right nightstand*). Go to the menu *Group* and choose *Open*. Click on the book to select it, right-click and choose *Isolate Selection*. Click on the *VRayCam001* viewport name, choose *Perspective*, and zoom in the book. Create a new *VRayMtl* and apply it to the book. Create a *VRayBitmap,* connect it to the *Diffuse map,* and load the image *kinfolk book_01* from the folder *Chapter 7*. The material does not apply correctly to the book and more specifically the pages of the book have the same texture as the cover (Figure 7.38).

In objects like this, where you want to apply two or more materials to the same object, you need to create a *Multi/Sub-Object material*. You have created the first material, which is the book cover, and now you need to create a second material, which is the book pages. Create a new *VRayMtl*, create a *VRayBitmap*, connect it to the *Diffuse map,* and load the image *pages*. Right-click in the *active View* and choose *Materials > General > Multi/Sub-Object* (Figure 7.39). Double-click on the *Multi/Sub-Object* node to see its parameters (Figure 7.40). A *Multi/Sub-Object* is a material that consists of multiple sub-materials. Each material is assigned an ID. By default, a *Multi/Sub-Object* has 10 sub-materials. In our example, we want the book to have two sub-materials, so click on the *Set Number* button, type 2, and press *OK* (Figure 7.41).

Connect the book cover *VRayMtl* to *ID* 1, the pages *VRayMtl* to *ID* 2, and apply the *Multi/Sub-Object* to the book (Figure 7.42).

If you check Figure 7.42, the book preview now looks correct, but let us see how you set the *IDs* to an object. Select the book, go to the *Modify* tab, to the *Selection* rollout and click on *Element*. If you click on the book cover, the full book is selected, so the *Element* command is not helpful for the specific object (Figure 7.43—left). In this case, you need to use the *Polygon* command and click on the book cover (Figure 7.43—right).

Go to the *Modify* tab and to the *Polygon: Material IDs* rollout. In this rollout, the *Select ID* field informs you that the material *ID 1* from the *Multi/Sub-Object* is assigned to the selected surface (Figure 7.44). Click on the *Select ID* button and all the surfaces that have the material *ID 1* will be selected (Figure 7.45).

Figure 7.38

Preview of book material.

7.1 The Scene Materials

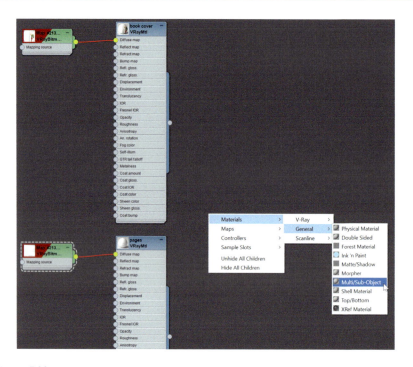

Figure 7.39
Steps to create a *Multi/Sub-Object*.

Figure 7.40
Preview of a Multi/Sub-Object.

Figure 7.41
Set Number of materials.

Figure 7.42
Preview of the book after applying the Multi/Sub-Object.

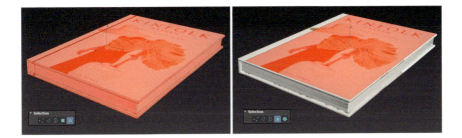

Figure 7.43
Element command selects the full book (left), while *Polygon* command selects specific polygons of the object (right).

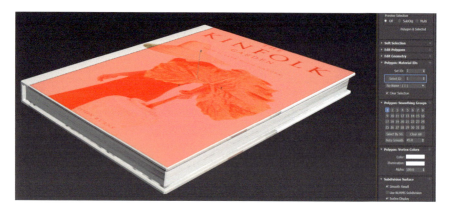

Figure 7.44

Material ID: 1 is assigned to the selected surface.

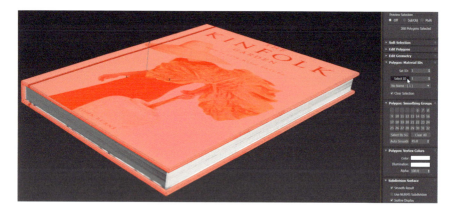

Figure 7.45

Select ID: 1.

Click on the pages and the *Select ID* field changes to 2 (Figure 7.46).

If you want the pages to get the *Material ID* 1, go to the *Set ID* field, type 1, and press *Enter*. This was just a test to see how you can change the *Material ID* of a surface. So, type again 2 and press *Enter* to assign the *Material ID* 2 to the pages.

When you finish adjusting the IDs of an object, you need to disable the *Polygon* command. Scroll up to the *Selection* rollout and click again on *Polygon* to disable it (Figure 7.47). Right-click in a viewport, choose *End Isolate,* and produce a render to see the book (Figure 7.48).

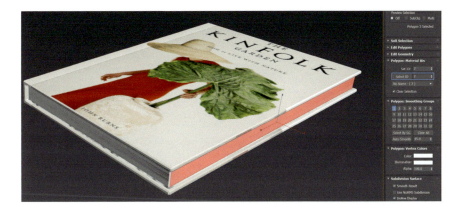

Figure 7.46

Material ID: 2 is assigned to the selected surface.

Figure 7.47

When an icon is highlighted, it is active.

Figure 7.48

Crop render of book.

7.1.9 Greenery (Multi/Sub-Object Material)

For the greenery in the vase, we will create two different materials. Select the objects *greenery type a* and *greenery type b* and right-click to isolate them. Moreover, turn to a *Perspective* view and zoom in to clearly see them (Figure 7.49).

> If when you click on the greenery, the object *decor right nightstand* is selected. This means that the group is closed. With the group selected, go to the menu *Group* and choose *Open*.

Figure 7.49
The flowers isolated in the *Perspective* viewport.

Create a *VRayMtl*, rename it to *greenery type a*, and assign it to the respective object. Create a *VRayBitmap*, connect it to the *Diffuse map*, and load the texture *branches* from the folder Chapter 7. Set the *Reflect color* RGB to 152 and the *Glossiness* to 0.85.

Regarding the object *greenery type b*, you need to create a *Multi/Sub-Object* with two material IDs; One for the leaves and the other for the branches. Thus, right-click in the *active View* and choose *Materials > General > Multi/Sub-Object*. Double-click on the node to open its parameters, click on the *Set Number* button, and type 2. Create a new *VRayMtl* and connect it to the material ID 1. In material ID 2 connect the *VRayMtl* that you created earlier for the *greenery type a* (Figure 7.50).

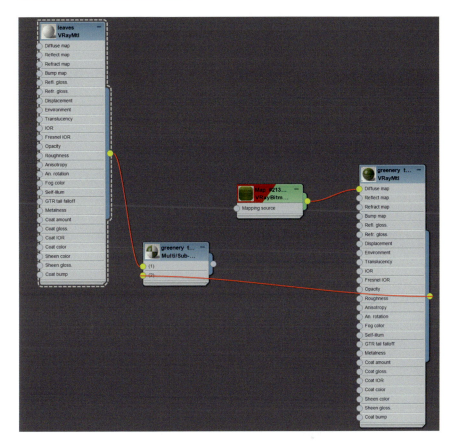

Figure 7.50
Steps to create the greenery type b material.

Create a *VRayBitmap* and connect it to the *Diffuse map* of the material ID 1. Load the image *greenery leaves*. Double-click on the *VRayMtl* node to open its parameters and set the *Reflect* RGB to 152 and the *Glossiness* to 0.85. Assign the *Multi/Sub-Object* to the *greenery type b* (Figure 7.51).

The next step is to define which polygons of the object should have ID 1 and which should have ID 2. Isolate the object *greenery type b*, go to the *Modify* tab, and enable the *Element* command from the *Selection* rollout. Either by clicking on the branches or by using the window selection, select the branches. Hold down the *Ctrl* key to add elements to your selection. Figure 7.52 shows the branches selected. Go to the *Modify* tab and to the *Polygon: Material IDs* rollout. Go to the *Set ID* field, type 2, and press *Enter*. The selected branches are assigned the

7.1 The Scene Materials 175

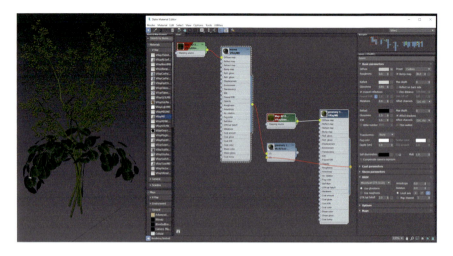

Figure 7.51

Multi/Sub-Object material applied to the *greenery type b*.

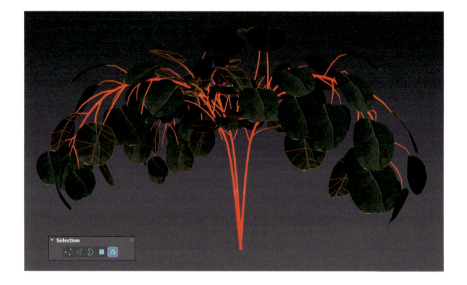

Figure 7.52

Branches selected using the *Element* command.

VRayMtl connected to the ID 2 (Figure 7.53). Produce a render to see the result (Figure 7.54).

Based on the above, you can assign materials to the remaining objects of the scene and get the result of Figure 7.55.

176　　　　　　　　　　　　　　　　　　　　　　　　　　　　7. Materials

Figure 7.53
Greenery type b after adjusting the IDs.

Figure 7.54
Crop render of the greenery.

7.1 The Scene Materials

Figure 7.55
Final render after applying all the materials.

7.2 Asset Tracking

3ds Max works with paths. When you create a material and you load a texture, you select a path for this texture. In the example we are working on, all the textures are in the folder *Chapter 7*. If you move any of these bitmap files (or the ies files) to another location, the scene will no longer render properly the respective material (or light). *Asset Tracking* allows you to re-path and relocate the missing files. As an example, go to the folder *Chapter 7*, select the image *leather*, cut it, and paste it to your desktop. Do a crop render at the bed base and you will see that it renders as gray leather. The attributes of the leather material are still there, i.e., the *Glossiness* and the *Bump* map. The only thing missing is the *Diffuse* map, where *V-Ray* uses the original, no-longer-valid path.

To access the *Asset Tracking* to fix the path, go to the menu *File* and choose *Reference* > *Asset Tracking Toggle* or press *Shift+T*. The *Asset Tracking* window opens (Figure 7.56).

The first thing you need to do is to click on the *Refresh* button, ▣ (Figure 7.57). If you forget to do so, then the list of maps seen in this window will not update.

Go to the *Status* column and check the row(s) with the message *File Missing*. Click on that row and go to the menu *Paths* and choose *Set Path...* (Figure 7.58). Click on the *3 dots* button and select the destination folder, i.e., the desktop.

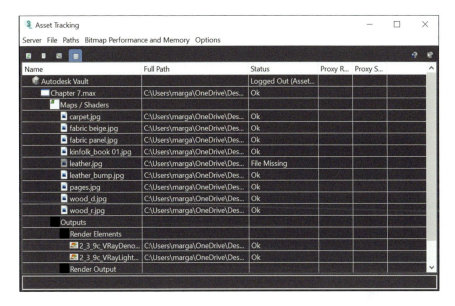

Figure 7.56
Asset Tracking window.

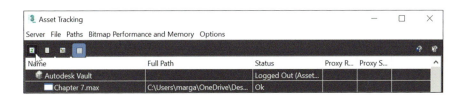

Figure 7.57
Refresh button.

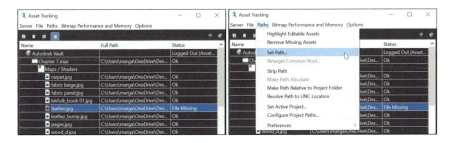

Figure 7.58
Steps to load the missing file.

7.2 Asset Tracking   179

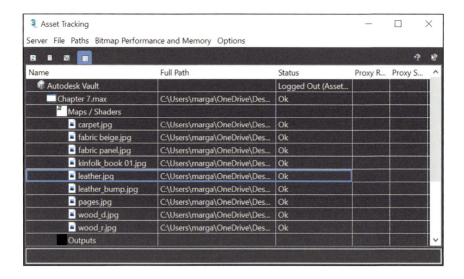

Figure 7.59
Updated path of the previously missing map.

Click *Use path* and press *OK*. If you now check the *Status* column, instead of *File Missing* the status changed to *OK*. Moreover, in the *Full Path* column, you see the updated path of the texture (Figure 7.59).

If you close the *Asset Tracking* window and produce a crop render, now the bed base renders properly.

Before you produce your final render, you should go to the *Asset Tracking* and check if there are missing textures, to confirm that your final image will render properly.

Do not forget to first press the *Refresh* button, every time you access the *Asset Tracking* window.

7.3 UVW Map

We have seen that to change the scale of a texture, you need to go to the *VRayBitmap* parameters, to the *Coordinates* rollout and adjust the *Tiling*. Another way is to use the *UVW Map* modifier. To examine this option, open a new *3ds Max* scene and create a box with *Length*: 20 cm, *Width*: 200 cm, and *Height*: 100 cm. Subsequently, open the *Material Editor* and apply a *VRayMtl*, which you can rename to *brick*. Add a *Diffuse* map that has a pattern, like a brick texture. For my projects, I mainly visit *textures.com,* which offers for free a great variety of textures. Try the *PBR Materials* (*Procedural Materials*). They are seamless textures

Figure 7.60

Brick Walls category in the *PBR Materials* menu.

offered as a package of textures that includes the *Diffuse map*, the *Reflect map*, the *Bump map*, etc., for every material.

Go to the *Brick Walls* category and click on the *Rustic Brick 2* preview (Figure 7.60). Scroll down to find the *Flat Maps* section. The *Albedo* is the *Diffuse map*, click to download it (Figure 7.61).

Figure 7.61

Download the *Albedo* map.

7.3 UVW Map 181

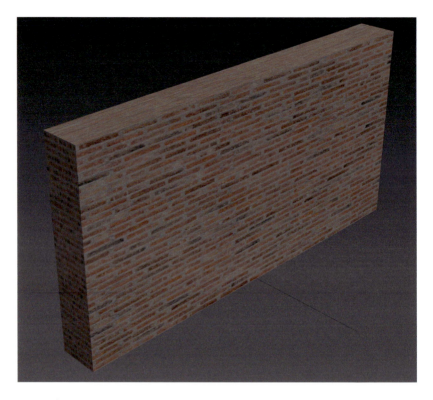

Figure 7.62

Preview of the brick texture on the box.

Create a *VRayBitmap* and load the *Albedo*. Then connect the *VRayBitmap* to the *Diffuse* map. Wire the *VRayMtl* to the box and do not forget to enable the *Show Shaded Materials in Viewport* command to see the texture in the *Perspective* viewport.

In Figure 7.62, it is seen that the texture does not apply in the same way on each side. The front view of the box looks good, but on the left side, the texture looks squeezed, while on the top side the texture looks stretched. If you observe each side more carefully, you will notice that the texture adjusts to fit each side. So, if you go to the box dimensions and change them, you will see that no matter the size of the box, the texture always stretches to fit the size of the box.

We mentioned in Chapter 6 that the size of a texture can be adjusted from the *Tiling* in the *Coordinates* rollout. So, double-click on the *Diffuse map* node, go to the *Coordinates,* and do some testing with the *U* and *V Tiling*. You will notice that no matter what value you enter, you still have the same problem, i.e., you cannot make the size of the brick equal on all sides.

In cases like this, you need to use the *UVW Map* modifier. In *3ds Max, U, V,* and *W* stand for the x, y, and z axes. A xyz axis system is placed at the bottom

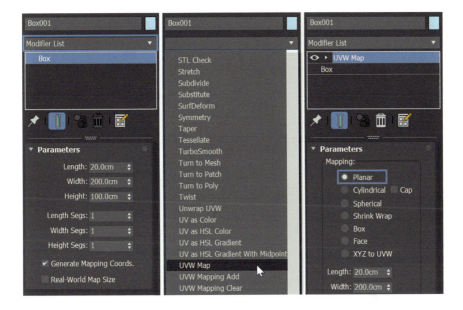

Figure 7.63

Steps to add the *UVW Map* modifier.

left corner of each viewport and when you select an object, another xyz system is located on the selected object.

Select the box, go to the *Modify* tab, click on the *Modifier list*, scroll down, and select *UVW Map* (Figure 7.63). With the default settings of the *UVW Map*, the texture on the box sides looks distorted (Figure 7.64). Go to the *UVW Map* parameters in the *Modify* tab and in the *Parameters* rollout there is a list of mapping methods. The default one is *Planar*.

Go through each option to see that the way the texture is projecting on the box depends on the mapping method (Figure 7.65).

When *Planar* is selected, you see an orange plane at the center of the box (Figure 7.65—left). The *Planar* mapping means that the texture will be projected on a flat plane. To fully understand this, go to the *Modify* tab and the *UVW Map* modifier. When there is an arrow next to a modifier, this means that the modifier has a submenu. Click on the arrow and choose *Gizmo*.

When *Gizmo* is disabled and you use the *Select and Move* command, the object moves. In contrast, when *Gizmo* is enabled and you use the *Select and Move* command, the object remains still whereas the texture of the object moves.

When *Gizmo* is enabled, you cannot select any other object in the scene. So, once you finish adjusting the *Gizmo*, do not forget to disable it.

7.3 UVW Map

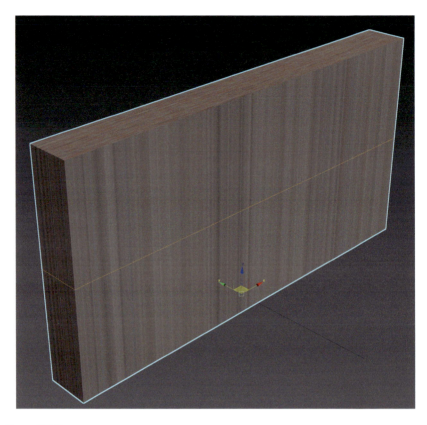

Figure 7.64
Box preview after applying the *UVW Map* modifier.

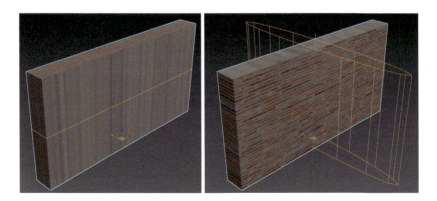

Figure 7.65
Planar mapping (left) and *Cylindrical* mapping (right).

Figure 7.66

Gizmo enabled.

When *Gizmo* is enabled, it gets a blue color in the *Modify* tab, and the xyz axes move to the center of the object, where this plane is. If you drag the cursor, the texture moves. Drag it outside the box, so that you can clearly see the planar shape (Figure 7.66).

With the *Gizmo* enabled, choose the *Select and Rotate* command, go to the *Coordinates* area and type 90 in the *X* field to rotate the texture by 90° around the x axis. The texture now applies accurately on the front and back view of the box, but not on the other sides (Figure 7.67).

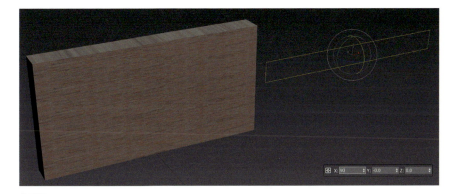

Figure 7.67

Box preview after rotating the texture by 90° around the x axis.

7.3 UVW Map

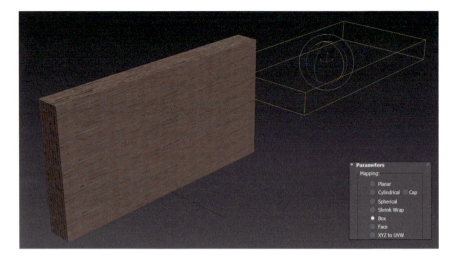

Figure 7.68
Box mapping method.

As mentioned earlier, the planar mapping means that the texture will be projected on a flat plane. Flat plane means that there are no sides, so no matter how you will rotate the plane, it can never project on all sides of the box equally well. To overcome this problem, go to the M*apping* options and choose *Box*. The reference shape changes to a box and the texture projects on all sides (Figure 7.68). But you still do not have the result you are looking for, i.e., to project the texture of the brick equally on all sides.

In order to obtain a realistic image of the wall, we need to know the size of the texture used. In our example, if we open the brick texture (*Albedo*), we can easily estimate that each side of this texture is 200 cm because it has 10 bricks per row, and if we assume that each brick is approximately 18 cm long plus the gaps in between, then each side seems like 200 cm. Now go to the *Parameters* rollout and below the *Mapping* options there are the *Length, Width,* and *Height* fields that define the size of the texture. When you first apply the *Box* mapping, these values match the size of the object. Set the *Length, Width*, and *Height* to 200 cm (Figure 7.69).

After setting the *Length, Width,* and *Height* to 200 cm, you can see in the *Perspective* viewport that the texture on each side of the box is equal to the other and the next step is to fix the orientation of the texture on the side of the box. With the *Gizmo* enabled and the *Select and Rotate* command active, go to the *Coordinates* fields and type 90 in the *X* field. The texture rotates by 90° and now the texture is projected correctly on all sides of the box (Figure 7.70).

Below the *Length, Width*, and *Height* fields, there are the *U, V,* and *W Tile* fields. They control the number of times the texture will repeat on a surface (Figure 7.71). Thus, if you make the *U Tile* value 2, you set the texture to be double

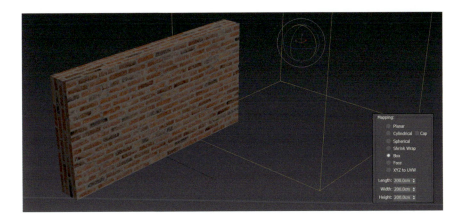

Figure 7.69
Box preview with *Length*, *Width*, and *Height*: 200 cm.

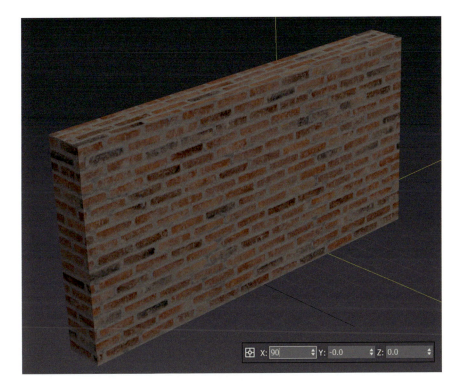

Figure 7.70
Texture rotated by 90°.

7.3 UVW Map

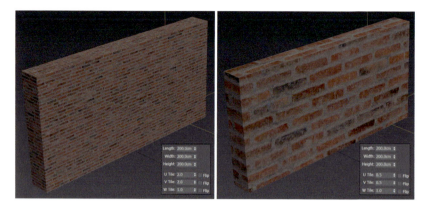

Figure 7.71

U and *V Tile*: 2 (left), *U* and *V Tile*: 0.5 (right).

dense on the x axis, while if you additionally make the *V Tile* value 2, you make the texture double dense on both axes. The opposite effect occurs if you use the value 0.5 for the *U Tile* and *V Tile*.

You can either use the *U, V,* and *W Tile* in the *Modify* tab or the *U, V,* and *W Tiling* fields in the *Coordinates* rollout in the *Material Editor* to control the tiling of a texture.

Once you set the correct scale of the texture, you can use the *Select and Move* command, to move the bricks if needed, so that the texture does not start or end with incomplete bricks (Figure 7.72).

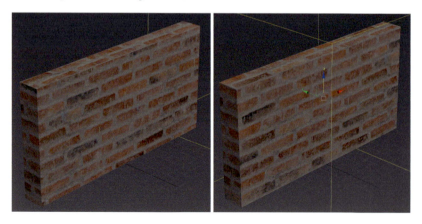

Figure 7.72

Before (left) and after (right) moving the texture.

Figure 7.73
Gizmo disabled.

Do not forget once you finish adjusting the texture to disable the *Gizmo*, otherwise, you will not be able to select other objects of the scene (Figure 7.73).

7.3 UVW Map 189

8
Libraries

V-Ray 5 comes with its own extensive libraries; two material libraries, the *VRayMtl Presets* and the *V-Ray Material Library*, and a 3D Models and HDRI library, the *Chaos Cosmos Browser*. These libraries can be a great starting point whether you are a beginner or an advanced user. You can use these pre-made materials and 3D Models as a base and then adjust them to fit your project needs. This way you achieve speed and realism.

8.1 VRayMtl Presets

Open the file *Chapter 8.max*, which contains the bedroom scene we were working on in the previous chapter. The *VRayPhysicalCamera* is set to *Exposure value* 13 and the *Exposure* in the *Exposure* layer is set to 4 (see Chapter 2).

The *VRayMtl Presets* is a drop-down menu with preset values for commonly used materials, like glass, chrome, gold, etc. To access the preset library, create a *VRayMtl*. Double-click on the node to open its settings and go to the *Basic parameters* rollout. Next to the *Diffuse* is the *Preset* setting. Click on *Custom* and a drop-down menu appears with pre-made materials to choose from (Figure 8.1). You can create different types of glass, metal, or fabric at the click of a button. You can also go to their parameters and adjust the settings and, for instance, convert a clear glass to a colored glass, as described in Chapter 7.

8.2 V-Ray Material Library

The *V-Ray Material Library* is a library with hundreds of high-quality *V-Ray* materials. It is introduced in *V-Ray 5*, so if you have a previous version of *V-Ray*, you will not have access to it. To open the *V-Ray Material Library*, go to the

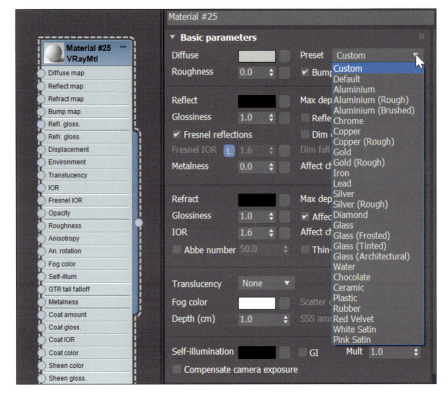

Figure 8.1

Preset drop-down menu.

V-Ray Toolbar and click on the *V-Ray Material Library Browser* button, (Figure 8.2).

If you do not see the *V-Ray Toolbar,* right-click on an empty spot at the ribbon of a toolbar and select the *V-Ray Toolbar.*

The *V-Ray Material Library Browser* opens (Figure 8.3). On the left side there is a list of material categories, like *Bricks, Concrete, Fabric,* etc., while on the right side there are previews of the materials each category includes. Open the *Material Editor* and resize both the *Material Editor* and the *V-Ray Material Library Browser* so that you can clearly see them both (Figure 8.4).

Figure 8.2

Part of *V-Ray Toolbar* and *V-Ray Material Library Browser* button.

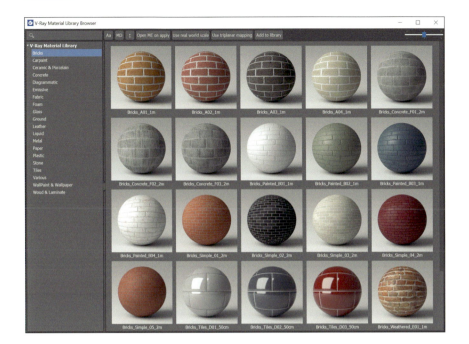

Figure 8.3

V-Ray Material Library Browser.

Figure 8.4

V-Ray Material Library Browser and *Material Editor* side by side.

8.2 V-Ray Material Library

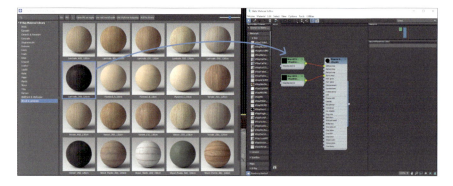

Figure 8.5
Use the drag and drop method to move a template material in the *Material Editor*.

From the *V-Ray Material Library Browser* go, for instance, to the *Wood & Laminate* category, select the *Laminate_D02_120cm* and drop it to the *Material Editor* (Figure 8.5).

Close the *V-Ray Material Library Browser* and assign the laminate material to the floor. If you click on the *Diffuse VRayBitmap* node and then on the *Show Shaded Material in Viewport* button, the floor appears with a gray color. Moreover, if you zoom in the material nodes, both the *Diffuse map* and the *Bump map* previews appear black (Figure 8.6). This means that the textures did not load.

To load a missing texture, double-click on the *VRayBitmap* node, then on the *3 dots* button in the *Parameters* rollout, and go and locate the missing texture (Figure 8.7).

The default location of the *Material Library* is *C:\Users\USERNAME\Documents\V-Ray Material Library\assets*.

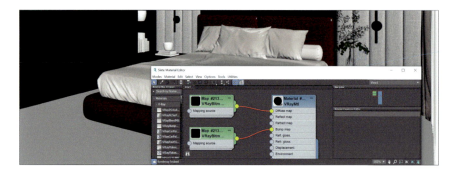

Figure 8.6
The material appears as a gray color in the camera viewport, while the previews in the *Material Editor* are black.

194   8. Libraries

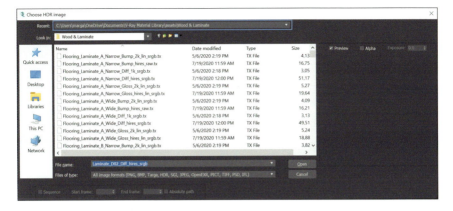

Figure 8.7

Path to load the missing textures.

After loading the textures, the material is visible in the camera viewport and in the *Material Editor* previews, and, therefore, you can produce a render (Figures 8.8 and 8.9).

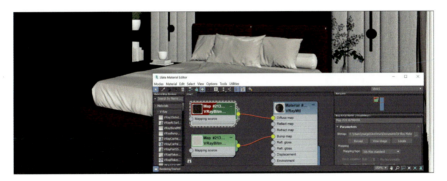

Figure 8.8

The floor material visible in the camera viewport and in the *Material Editor* previews.

Figure 8.9

Crop render of the floor.

8.2 V-Ray Material Library 195

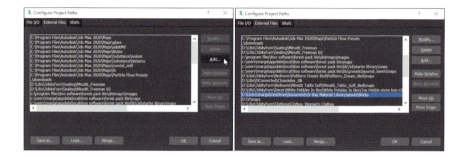

Figure 8.10

Steps to add external files.

In the scene, you might want to test several of the materials in the library until you reach to the final result. Instead of having to link every missing bitmap each time, you can set in *3ds Max* where to look for the library assets. To do so, go to *File > Project > Configure Project Paths…* Select the *External Files* tab and click on the *Add* button (Figure 8.10). Go to the location of the *Material Library* folder (*C:\ Users\USERNAME\Documents\V-Ray Material Library\asset*), where you need to choose and add each category, i.e., *Bricks, Carpaint, Ceramic & Porcelain*, etc., and press *OK*.

Another way to link textures is by using the *Asset Tracking* (see Chapter 7).

After setting the path, go and test more materials. Open the *V-Ray Material Library Browser*, go to the *Leather* category, click on the *Leather_B01_ Brown_25cm*, and drag it to the *Material Editor* (Figure 8.11). Apply the leather to the bed base and produce a crop render (Figure 8.12).

Figure 8.11

The drag and drop method applied to create a leather.

Figure 8.12
Crop render of the bed with the new leather applied.

If you check the *Material Editor*, the preview of the *VRayBitmaps* appears black, but the material renders properly. Once you re-start the software, the previews of the materials will look right.

8.3 Cosmos Browser

Apart from the material libraries, *V-Ray 5* offers a 3D models library as well. Go to the *V-Ray Toolbar* and click on the *Cosmos browser* button, ◉ (Figure 8.13). The *Chaos Cosmos Browser* opens (Figure 8.14).

On the left side there are the menus *Home, 3D Models, HDRIs,* and *Creators*. We explored the *HDRIs* menu in Chapter 4, and in this chapter, we will analyze the *3D Models* menu. The *Creators* menu is basically a combination of the *3D Models* and the *HDRIs*. Click on the *3D Models* menu and a sub-menu appears organizing the 3D models into categories. Choose *Accessories > Tableware &food*. When you leave the cursor over a preview, you see the *Download* button, where you can click to download the respective model (Figure 8.15). If you click on a preview, the properties of the object open. Click on *Glass 004*. This 3D model is offered by *Design Connected* and it is approximately 14 MB. On the left side you

Figure 8.13
Part of *V-Ray Toolbar* and *Cosmos browser* button.

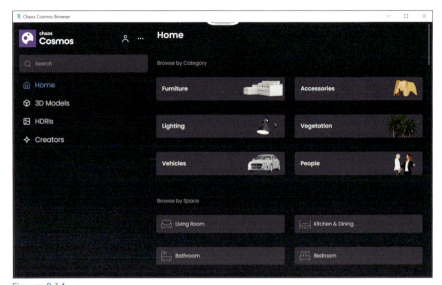

Figure 8.14

Chaos Cosmos Browser.

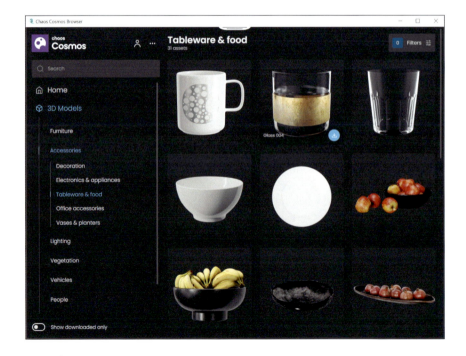

Figure 8.15

Tableware & food category of the Chaos *Cosmos Browser.*

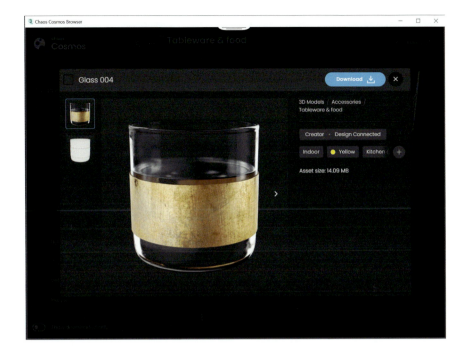

Figure 8.16

Glass 004 properties.

can choose to see the shaded or the wireframe preview of the selected model. Click on the *Download* button to download it (Figure 8.16).

> You need to download a 3D Model only once. The *Glass 004* is now in your library and you do not need to download it again to use it in future projects.

To add a 3D model to the scene, simply drag and drop it in. Go to the *Top* viewport, click on the *Glass 004*, and drop it on the book on the right nightstand (Figure 8.17). Isolate the *Glass 004* and the nightstand accessories and move the glass to the correct position both in the *Top* and in the *Front* viewports (Figure 8.18).

Do a crop render at the glass (Figure 8.19).

> Do not forget that when you do a crop render, you can increase the resolution to see that part of the scene better.

If you want to change the finishes of the object, open the *Material Editor*, click on the *Pick Material from Object* command, , and click on the object, *Glass 004* (Figure 8.20).

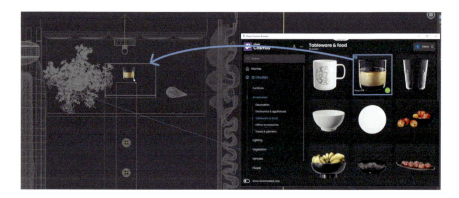

Figure 8.17
The drag and drop method applied to add the *Glass004* in the scene.

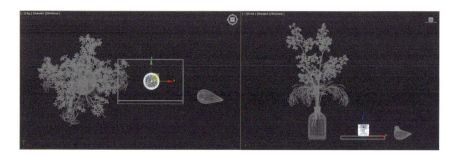

Figure 8.18
Repositioning of the *Glass004*.

Figure 8.19
Crop render of the *Glass004*.

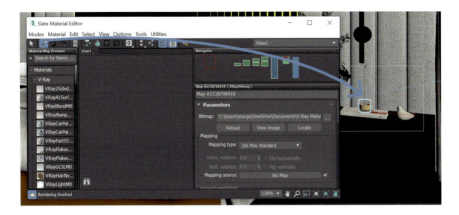

Figure 8.20

Use of the *Pick Material from Object* command on *Glass 004*.

The object's material is transferred to the *Material Editor* and you can adjust it. *Glass 004* is assigned a *Multi/Sub-Object* that consists of three sub-materials, a glass material with ID 1, a copper material with ID 2, and a water material with ID 3 (see Chapter 7). If, for instance, you want to make the glass material the same glass as the one applied to the vase on the nightstand, use the *Pick Material from Object* command and click on the vase to transfer that material in the *Material Editor* as well. Then connect the *glass vase VRayMtl* to *(1)* of the *Multi/Sub-Object* (Figure 8.21). Produce a render (Figure 8.22).

8.4 Bedroom Transformation

Using the three libraries described above, we can change the materials and accessories to give the bedroom a new look. For example, we can apply to the floor the *Flooring_Laminate_Wide_250cm* from the *V-Ray Material Library* (Figure 8.23).

For the wardrobe, the nightstands, and the slats, you can create a new *VRayMtl*, choose the *Plastic* preset, make the *Diffuse color* white, and the *Reflect Glossiness* 0.75 to give the feeling of a white matte lacquer finish (Figure 8.24).

Finally, regarding the materials, go to the settings of the metal you have already created and applied to the wall sconces and the ceiling tracks, and make the *Diffuse color* a white one. In what concerns the accessories, you can delete the decorative sets on both nightstands, the plant, and the black books in the niche and put in their place the following: On the left nightstand, the *Oranges 001* from the *Tableware & food* category. On the right nightstand, the *Potted Succulent 002* from the *Indoor plants* category in combination with the *Books 007* from the *Decoration* category. On the shelves, you can use the *Indoor Plant 006* and scale it down approximately at 35% to fit in and finally choose the *Books 001* (Figure 8.25). The render you will get should look like the one in Figure 8.26.

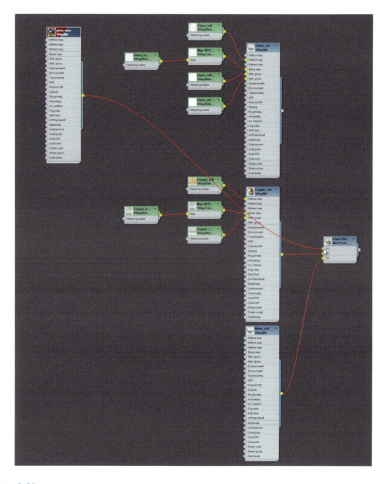

Figure 8.21
Connecting the *glass vase VRayMtl* to ID:1 of the *Multi/Sub-Object* material.

Figure 8.22
Crop render of the new glass on the *Glass 004*.

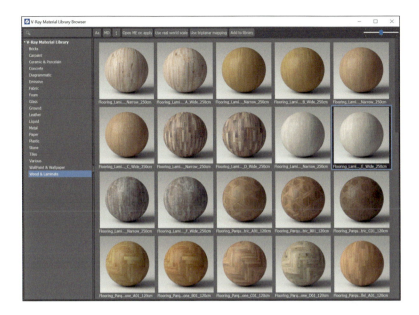

Figure 8.23
New selection for the bedroom flooring.

Figure 8.24
New *VRayMtl* using the *Plastic* preset for the wardrobe, nightstands, and slats.

8.4 Bedroom Transformation

Figure 8.25

New accessories selections.

Figure 8.26

Bedroom render after changing the materials and the accessories.

You can place more cameras and set close-up views to see the new accessories and materials. For example, you can set a camera at the right side of the right nightstand to capture the new books and the succulent, as seen in Figure 8.27. Do not forget to use the *Clipping planes* to move a camera behind an object and see beyond that object. Figure 8.28 shows the new render produced.

Another camera you can set is in front of the left nightstand. In this example, you can place the Camera higher than the Camera Target to have a more artistic result (Figures 8.29 and 8.30).

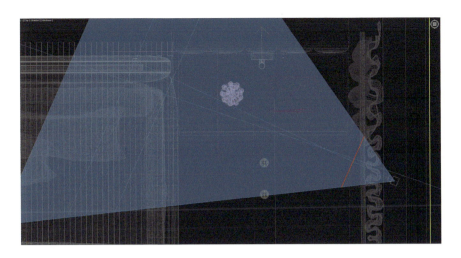

Figure 8.27
New camera placement at the right nightstand.

Figure 8.28
Close-up render of the accessories on the right nightstand.

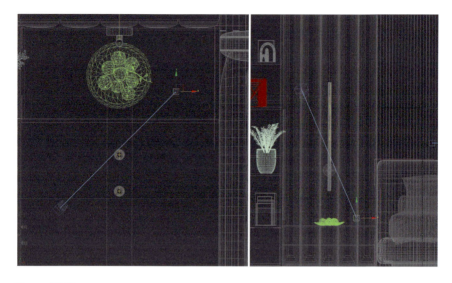

Figure 8.29
New camera placement at the left nightstand.

Figure 8.30
Close-up render of the accessories on the left nightstand.

Index

3D Models, 197
3ds Max interface, 1

A

Absolute Mode Transform Type-In, 12
Antialiasing, 39
Archive, 16
Area to Render, 51, 105, 151
Asset Tracking, 178, 196
Assign Material to Selection, 133, 138, 153
Assign V-Ray, 19

B

Bloom/glare effect, 29, 111
Brute Force, 49
Bucket Image sampler, 22, 45
Buckets, 42. 45
Bump, 142, 163
Bump map, 142, 163

C

Cameras, 61
    Camera Lister, 83
    Camera Viewport Controls, 72
    Selecting a Camera, 63
    Simulation of Real-Life Cameras, 73
    Sensor & Lens, 68
    Show Safe Frames, 54, 63
    Truck Camera, 72

VRayPhysicalCamera, 61, 73
VRayPhysicalCamera
    Settings, 67
    Aperture, 74, 79
    Clipping, 68
    Depth of Field, 61, 75, 79, 83
    F-Number, 78
    Field of View, 61, 68, 80
    Film speed (ISO), 74, 78
    Focal length, 68
    Exposure value (EV), 74
    Shutter speed, 74
Clipping, 68
    Near clipping plane, 70
    Far clipping plane, 70
Clone 104
    Copy, 83, 104, 142
    Instance, 104, 111, 123
Command Panel, 2, 8
    Create, 2, 8, 10, 61, 85, 96, 102, 116
    Modify, 2, 9, 68, 76, 95, 106, 175, 183
    Utilities, 2, 15
Compact Material Editor, 127, 138, 143
Contrast, 24, 26, 101
Coordinates Display, 12, 67
Create layer, 24, 26, 27
Create New View, 138
Coordinates, 3, 12, 67, 87, 136, 180

Cosmos browser, 197
   3D Models, 197
   Creators, 197
   HDRIs, 91, 197
Curves, 27

D

Depth of Field (DoF), 61, 75, 79, 83
Diffuse, 133, 139, 149, 151, 157, 159, 161, 163, 168, 169, 201
Dimensions, 9, 15, 151
DoF & Motion blur, 79
Dolly Camera + Target, 72
Dome VRayLight, 85, 92, 95
Double-sided, 107

E

Exposure, 24, 76, 78
Exposure value (EV), 74, 76

F

F-Number, 78
Field of View, 61, 68, 80
Film speed (ISO), 74, 78
Focal length, 68

G

GI, 37, 49
Gizmo, 183
Glass, 157, 191, 197
Global Illumination, 49
Global Switches, 37
Greenery 173
Go to Parent, 140

H

HDRI 91, 197
Hidden lights, 37
Hide selection, 33, 37, 58
Highlight Burn, 24, 76, 88, 101, 147
History, 23, 32
   A/B comparison, 35
   Save to history, 33

I

Image Sampler, 22, 39, 42, 45
Indirect illumination, 49

Instance, 104, 111, 123
Interactive Production Rendering (IPR), 57

L

Layers, 23, 93, 123
Lens Effects, 29, 111, 124
Libraries, 191
   Cosmos browser, 91, 197
   V-Ray Material Library, 191, 201
   VRayMtl Presets, 191
Light Cache, 50

M

Mapping, 183
Material Editor, 127
   active View, 4, 7, 127
   Basic parameters, 133, 157, 191
      Diffuse, 133, 139, 149, 151, 157, 159, 161, 163, 168, 169, 201
      Glossiness, 153
      Reflect, 153, 159, 167, 174
      Refract, 157, 161
   Compact Material Editor, 127, 138, 143
   Coordinates rollout, 136, 153, 164, 180
   Maps rollout, 142, 164
   Material/Map Browser, 127, 134, 137, 139
   Multi/Sub-Object, 169, 173
   Parameter Editor, 127, 133, 157
   Presets, 191
   Reset Mtl/Map Params, 143
   Show Shaded Material in Viewport, 135, 140, 149, 151, 156, 163
   Slate Material Editor, 127, 144
   Tiling, 136, 164, 180, 188
   VRayBitmap, 98, 134, 140, 149, 151, 161, 163, 168, 169, 180
   VRayMtl, 129, 138
Materials, 147
   Book, 169
   Carpet, 168
   Drapery, 161
   Fabric, 149
   Glass, 157
   Greenery, 173
   Leather, 163
   Metal, 159

Sheer, 161
Wood, 151
Maximize Viewport Toggle, 4
Measure, 15
Multi/Sub-Object 169, 173

N

Navigation Controls, 3, 4, 9, 72, 104
Noise threshold 44, 46, 120

O

Object Color, 127
Object Type, 8, 61, 102, 116
Offset Mode Transform Type-In, 12
Orbit, 5
Output Size, 52, 55

P

Physical exposure, 78
Pick Material from Object, 143, 199, 201
Pick mesh, 113
Presets, 191
Progressive Image sampler, 22, 42, 120

R

Refract, 157, 161
Region render, 59, 105
Reflect, 153, 159, 167, 174
Render Elements, 42, 46, 50, 92, 124
Render Output, 56
Render Production, 21
Render Setup, 19, 30, 37, 42, 46, 51, 57, 105, 111, 120, 151
Rendering Settings, 19, 37
 Assign V-Ray, 19
 Common, 37, 51
  Area to Render, 51, 105, 151
  Output Size, 52, 55
 GI, 37, 49
  Brute Force, 49
  Global illumination, 49
  Light Cache, 50
 Image sampler, 22, 39, 42, 45, 120
  Antialiasing, 39
  Bucket Image sampler, 22, 45
  Progressive Image sampler, 22, 42, 120
  Max. subdivs, 42, 46

  Min. subdivs, 42, 46
  Noise threshold, 44, 46, 120
  Render time, 42, 87, 120
 Render Elements, 42, 44, 46, 50, 92, 124
  VRayDenoiser, 42, 44, 46, 50, 124, 160
  VRayLightMix, 50, 93, 123
 Rendering window 21, 22
 Save current channel 32, 124
 V-Ray 37
  Global Switches, 37
 VFB, 20, 23
 V-Ray Frame Buffer, 20
  History, 23, 32
  Layers, 23, 93, 123
  V-Ray Messages, 21, 23
Reset Mtl/Map Params, 143

S

Save, 16, 32, 56
Save As, 16
Save Selected, 16
Select and Move, 12, 65, 87
Select and Rotate, 13, 95
Select and Scale, 14
Select an Object, 11
Select by Name, 11, 112
Select from Scene, 11, 63, 113, 151
Select Camera, 62, 64
Select Camera Target, 64
Selection Filter, 65, 111
Show Safe Frames, 54, 63
Show Shaded Material in Viewport, 135, 140
Shutter speed, 74, 76, 78, 79
Slate Material Editor, 127, 144
Standard Primitives, 8
Status bar, 3, 12, 67, 87
System Unit Setup, 8

T

Tiling, 136, 180, 188
Transform commands, 12, 65
 Select and Move, 12, 65, 87
 Select and Rotate, 13, 95
 Select and Scale, 14
Truck Camera, 72

U

Units Setup, 8
UVW Map, 180

V

V-Ray Camera Lister 83
V-Ray Frame Buffer (VFB) 20, 23
  History, 23, 32
  Layers, 23, 93, 123
  Save current channel, 32, 124
V-Ray Material Library, 191
V-Ray Toolbar, 3, 83, 91, 192, 197
VFB, 20, 23
ViewCube, 5
Viewport Configuration, 5
Viewport Navigation Controls, 3, 4
Viewports, 3, 4
  General, 5, 7
  Layout, 4, 5, 7
  Maximize Viewport Toggle, 4
  Per View Preference, 5, 7
  Point-of-View (POV), 5, 7
  Reset Layout, 4
  Shading, 5, 7
VRayBitmap, 98, 134, 140, 142

VRayDenoiser, 42, 44, 46, 50, 124, 160
VRayIES, 116
VRayLight, 91, 95, 101
  Double-sided, 107
  Intensity, 107
  Invisible, 107
  Temperature, 106
VRayLightMix, 50, 93, 123
VRayMtl, 129
VRayMtl Presets, 191
VRayPhysicalCamera, 61, 67, 73
VRaySky, 85
VRaySun, 85
  Placing VRaySun, 85
  VRaySky, 85
  VRaySun parameters, 88

W

Welcome Screen, 1
White Balance, 26, 122, 147
Whiteness, 133, 153

Z

Zoom Extends All, 10
Zoom Extends All Selected, 9, 103